AF248769

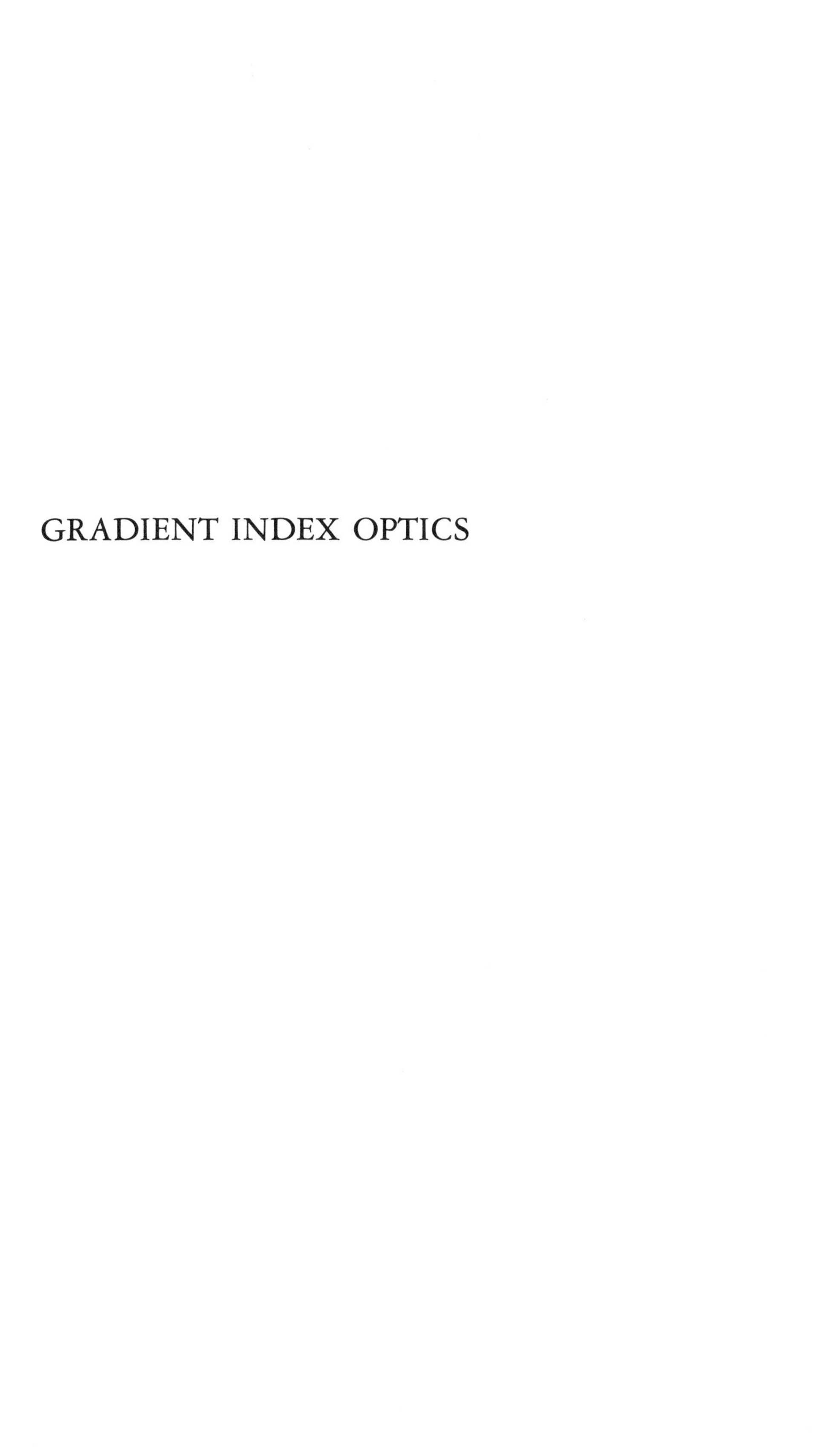

GRADIENT INDEX OPTICS

To Dorothy

DISCARDED

QC
285.2
.D77
M37

GRADIENT INDEX OPTICS

Erich W. Marchand

Physics Division
Research Laboratories
Eastman Kodak Company
Rochester, New York

1978

ACADEMIC PRESS   *New York   San Francisco   London*
A Subsidiary of Harcourt Brace Jovanovich, Publishers

LIBRARY
Appalachian State University
Boone, North Carolina 28608

Copyright © 1978, by Academic Press, Inc.
ALL RIGHTS RESERVED.
NO PART OF THIS PUBLICATION MAY BE REPRODUCED OR
TRANSMITTED IN ANY FORM OR BY ANY MEANS, ELECTRONIC
OR MECHANICAL, INCLUDING PHOTOCOPY, RECORDING, OR ANY
INFORMATION STORAGE AND RETRIEVAL SYSTEM, WITHOUT
PERMISSION IN WRITING FROM THE PUBLISHER.

ACADEMIC PRESS, INC.
111 Fifth Avenue, New York, New York 10003

United Kingdom Edition published by
ACADEMIC PRESS, INC. (LONDON) LTD.
24/28 Oval Road, London NW1   7DX

Library of Congress Cataloging in Publication Data

Marchand, Erich W
    Gradient index optics.

    Bibliography: p.
    Includes index.
    1. Lenses——Design and construction. 2. Optics,
Geometrical. I. Title.
QC385.2.D47M37   681'.42   78–11745
ISBN 0–12–470750–5

PRINTED IN THE UNITED STATES OF AMERICA

CONTENTS

PREFACE

An inhomogeneous medium is one in which the refractive index varies from point to point within the medium. Currently the terms "gradient index" and "graded index" are often used to describe such media. In the present book only isotropic media are considered, these being ones for which the refractive index at each point is independent of direction.

Inhomogeneous media occur frequently in nature. Examples are the lens of the human eye and the atmosphere of the earth. The possibility of using such media in man-made optical instruments has been considered for many years, but only recently has it been possible to fabricate the appropriate materials to exploit this idea.

At the present time, much research has been devoted not only to theoretical developments in this field, but also to improved methods of producing and measuring the necessary materials.

In view of the impressive advances in the field of homogeneous fiber optics within the past 20 years, it is natural that the idea of making gradient index fibers should be actively explored. This has, indeed, taken place, and already endoscopes made for the medical profession by this means are commercially available. The chief advantage of gradient fiber in this application is that a single fiber instead of a bundle of fibers can be used to transmit an image.

The communication industry has been developing the use of gradient index fibers as a highly efficient means of transporting information. Light losses in a gradient fiber tend to be very low, since rays within a fiber are curved in such a way that they are not reflected at the surfaces of the fiber.

Interesting as these applications are, they are not the chief subject of this book. There has already appeared an extensive literature devoted to gradient index optics, most of it devoted to the above two topics, image relays, and waveguides. Accordingly, we do not attempt to deal to any great extent with the fiber optics area, but concentrate instead on the application of gradients in optical systems of classical types: gradient index lenses.

It has been clearly shown mathematically that improved results can be achieved by gradients in this application, either through better performance, reduction of the number of lens elements, savings in weight and space, or in other advantages. Although gradient elements are still not easy to make, it is expected that they may result in cost savings in some cases, perhaps, for example, where aspheric surfaces can be replaced by gradient elements.

With respect to pure theory, a large body of literature already exists. The work of Luneburg (1964) contains a thorough development of the mathematical theory of inhomogeneous media, derived from first principles. It is not our intention to duplicate that material here. However, a certain amount of this basic theory must be reviewed in the present book in order to unify the subject matter.

A few years ago a survey article on gradient index lenses appeared (Marchand, 1973), but, since that time, a considerable body of literature has appeared dealing with both theoretical and technological advances in the field. The purpose of this book is therefore twofold: (1) to describe, partly in detail and partly in summary, the present state of theory and practice related to gradient index lenses, and (2) to identify many of the sources of information related to this field. It is expected that merely a substantial list of references may be helpful to workers who wish to become familiar with a new scientific area. In this connection a number of references and the related discussions apply primarily to the fiber optics field, since often the subject matter overlaps, both in theory and technology, the field of gradient index lenses.

With respect to notation, little or no attempt has been made to retain uniform symbols throughout the book. For example, r is used in one chapter as the distance from a point, and elsewhere as the distance from a line. However, a perceptive reader should have no difficulty with changes of this kind.

My sincere thanks go to David Hamblen, a member of my own department at Kodak, for helping me to understand some of the methods of making and measuring gradients. Also, occasional conversations with Duncan Moore of the Institute of Optics at the University of Rochester have been helpful. Special thanks go to the Eastman Kodak Company, who not only gave me time to prepare this book, but also put their many facilities at my disposal.

Chapter 1
HISTORICAL INTRODUCTION

1.1 Gradients in Astronomy

Studies of astronomical refraction date back to Cleomedes (100 A.D.) and Ptolemy (200 A.D.) as reported by Mahan (1962). Alhazan (1100 A.D.) already suggested that astronomical refraction was responsible for the flattening of the sun's disk near the horizon.

In 1587 Tycho Brahe made direct measurements of the magnitude of the refraction. However, the correct theoretical analysis of this phenomenon was not possible at that time since the exact refraction law was not yet known. A step in this direction was taken by Kepler, who concluded that a correction term should be added to Ptolemy's formula.

Cassini, in 1656, applied the correct Snell's law to the problem but assumed, as did the earlier investigators, that the atmosphere is homogeneous up to a fixed altitude. Later studies took better account of the inhomogeneity of the atmosphere by assuming it to be made up of concentric homogeneous shells. More recently, mathematical models have been proposed on the basis of a continuously varying refraction index having spherical symmetry about the center of the earth.

1

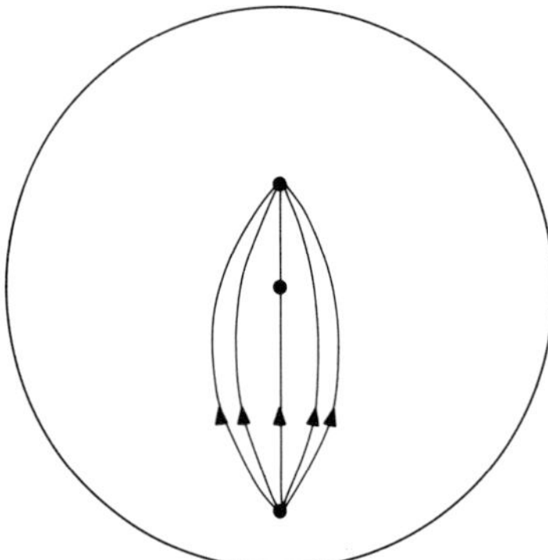

Fig. 1.1. Maxwell's fisheye lens.

1.2   Gradient Index Lenses

The possibility of using inhomogeneous media in optical systems has been considered for many years. Maxwell (1854) demonstrated that a medium having a suitable refractive index distribution can provide a nontrivial example of an absolute instrument, that is, one in which every point of a region of space is sharply imaged. The lens that he described, known as Maxwell's fisheye, involves an index function having spherical symmetry about a point and would be expected to have the form of a sphere (see Fig. 1.1).

The fisheye lens has remained a theoretical curiosity, as it is almost impossible to make and has little chance of serving any useful purpose. Only points on the surface and within the lens are sharply imaged. Furthermore, the images of extended objects suffer from severe aberrations. It is even doubtful whether any fish has an eye of this kind.

Luneburg (1964) discovered a lens that focuses every bundle of parallel rays into a point. This lens cannot be considered an absolute instrument since the points at infinity must be regarded as lying on a surface rather than filling a region of space.

The Luneburg lens also has an index function with spherical symmetry about a point. This lens, likewise, is difficult to make (at least for light in the visible region of the spectrum), since the index at its spherical surface must match that of the surrounding medium. However, any parallel bundle of rays incident on the lens passes through the lens and converges at a point located on the opposite surface of the lens (see Fig. 1.2). Thus, a Luneburg lens, if it can be made, has limited

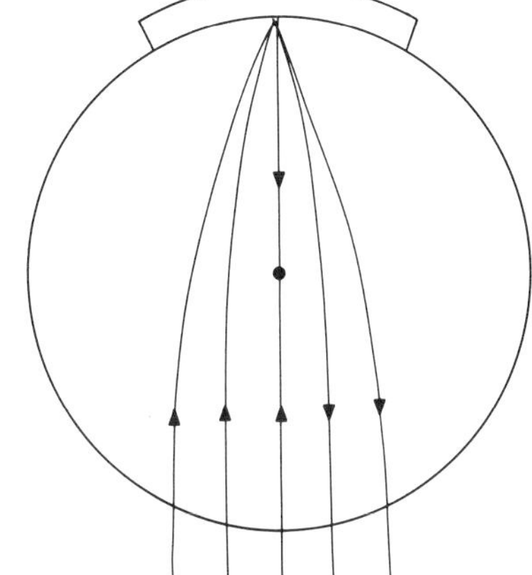

Fig. 1.2. Luneburg lens.

possibilities for useful application, although more so than the Maxwell fisheye.

An interesting modification of the Luneburg lens is obtained by forming a mirror on the image side of the sphere, as shown in Fig. 1.2. Then each entering ray, after reaching the mirror surface, will be reflected back out of the lens in a direction parallel but opposite to its original direction. This action is similar to that of the well-known corner cube consisting of three orthogonal plane mirrors.

Luneburg actually reported his lens prior to 1944. Morgan (1958) and others generalized Luneburg's formula and showed that a gradient index lens can be designed to image points of an external sphere sharply onto another external sphere. It is still impractical to fabricate such a lens for use with visible light. However, for microwave applications, it is possible to simulate a variable index of refraction at microwave frequencies by means of artificial dielectrics.

Wood (1905) devised a method of constructing a simple lens having two plane surfaces and a refractive index varying with the distance from the optical axis. His procedure was based on a dipping technique whereby a cylinder of gelatin is produced with an internal refractive index gradient having symmetry about the axis. By slicing the cylinder with plane cuts perpendicular to the axis, a number of disks are obtained each having plane faces and a radial index distribution. Wood showed that a disk of this kind, despite having only plane faces, acts like a converging or diverging lens depending on whether the index is a decreasing or increasing function of the radial distance.

1.3 Recent Developments

Because it seemed impossible to create practical gradient index
lenses, there was little interest in this subject until just a few years ago.
Actually, for some time glass technologists had experimented with the
effects of diffusing ions into glass in order to modify its refractive index.
It was finally discovered that this technique can be successfully applied
to produce substantial index gradients in certain types of glass while
leaving the glass strain free and colorless (see Hamblen, 1969; Pearson
et al., 1969).

The result of this breakthrough has been a renewed interest in the
subject of inhomogeneous media not only with respect to theoretical
implications but also with the hope of finding practical applications
of the new technology. Much effort has accordingly been expended
on the use of gradients as an extension of the already well-developed
science of fiber optics.

As for the utilization of gradient elements in optical systems of
classical types, such as microscopes, telescopes, and photographic
lenses, much theoretical work has also been done. In a series of papers
Marchand (1970, 1972, 1973, 1976), Sands (1970, 1971a,b,c,d), Moore
and Sands (1971), D. T. Moore (1971), and others have described the
mathematical theory pertaining to the use of gradients in optical
systems.

Although these investigations indicate clearly that gradients offer
the lens designer a powerful tool for creating new and improved
optical systems, the practical realization of the new theory has been
somewhat slow to arrive. The chief drawback seems to be that the
techniques for producing good-quality gradient materials at a reasonable cost have not advanced rapidly enough to keep pace with the
theoretical developments.

1.4 Basic Theory

In an inhomogeneous isotropic medium the refractive index is a
function

$$n = f(x, y, z) \tag{1.1}$$

of the coordinates of the points of the region being considered. The
question of determining the paths (usually curved) of the various

possible rays in such a medium was answered, in principle, long ago. The result can be expressed in the form of a single vectorial differential equation of second order, an equation which can be deduced easily from Fermat's principle. This well-known principle states that, if C is a ray joining any two points of the medium, the light-path integral

$$L = \int_{s_0}^{s} n \, ds \tag{1.2}$$

taken along C from the first point to the second is stationary relative to its value for any nearby curve joining the two points. Here s is the arc length along the curve.

Equation (1.2) can be used in a number of ways. If a particular coordinate system seems most appropriate for a particular problem, the integral can be expressed in this coordinate system, and the corresponding differential equations of the rays for that system are given by the familiar Euler equations from the calculus of variations. In using this method a judicious choice of the parameter to be used as variable of integration will yield the most convenient form for the resulting differential equations. Examples of this method will appear in later chapters.

If Cartesian coordinates are selected, either x or y or z can be chosen as the variable of integration. However, a common technique in such variation problems is the alternative of retaining s as an extra parameter, to be used as the variable of integration while introducing the dependence of the integrand on the curve through the relation

$$x'^2 + y'^2 + z'^2 = 1, \tag{1.3}$$

where a prime indicates d/ds.

This procedure preserves the formal symmetry with respect to x, y, and z. It is known that the corresponding four-dimensional variation problem will lead to the correct differential equations provided the integral is cast into the form

$$L = \int_{s_0}^{s} F(x, y, z, s, x', y', z') \, ds, \tag{1.4}$$

where it is required that F be a homogeneous function of first order with respect to x', y', and z'. Accordingly, with the help of Eq. (1.3), we express F in the form

$$F = n(x, y, z)(x'^2 + y'^2 + z'^2)^{1/2}. \tag{1.5}$$

The Euler equations now are

$$\left(\frac{\partial F}{\partial x'}\right)' = \frac{\partial F}{\partial x},$$

$$\left(\frac{\partial F}{\partial y'}\right)' = \frac{\partial F}{\partial y}, \tag{1.6}$$

$$\left(\frac{\partial F}{\partial z'}\right)' = \frac{\partial F}{\partial z}.$$

The vectorial form of these equations is

$$(n\mathbf{r}')' = \mathbf{\nabla} n, \tag{1.7}$$

where $\mathbf{r}$ is the position vector with components (x, y, z) and $\mathbf{\nabla} n$ is the gradient of n.

Equation (1.7) can be derived in other ways and is a well-known basic formula for determining the possible rays that can pass through a given medium specified by the index function n. This differential equation applies to any isotropic medium provided the function n is twice differentiable. If it is appropriate to introduce a different coordinate system, the corresponding differential equations can be obtained either directly from Eq. (1.2), as mentioned above, or by converting Eq. (1.7) to the new coordinates. In either case the differential equations of the rays will assume various forms depending on the choice of the independent variable.

Recently Moore (1975) has found that the differential equations of the rays can be put into a simple form in Cartesian coordinates if x or y or z is taken as the independent variable. For example, with z independent, we have

$$n\ddot{x} = (1 + \dot{x}^2 + \dot{y}^2)\left(\frac{\partial n}{\partial x} - \dot{x}\frac{\partial n}{\partial z}\right),$$

$$n\ddot{y} = (1 + \dot{x}^2 + \dot{y}^2)\left(\frac{\partial n}{\partial y} - \dot{y}\frac{\partial n}{\partial z}\right), \tag{1.8}$$

where a dot means d/dz. Here no assumptions are made on the symmetry of the index function. The derivation of these equations is given in Appendix A.

Chapter 2
SPHERICAL GRADIENTS

2.1 Introduction

A spherical gradient is a refractive index distribution for which Eq. (1.1) takes the form

$$n = f(r), \tag{2.1}$$

where

$$r = (x^2 + y^2 + z^2)^{1/2} \tag{2.2}$$

for a suitable choice of origin. In such a medium the index function has spherical symmetry about the origin.

It is easy to show that every ray in a spherical medium is a plane curve lying, in fact, in a plane through the center of symmetry. From Eqs. (1.7), (2.1), and (2.2) it follows that

$$n\mathbf{r}'' + n'\mathbf{r}' = \dot{n}\,\frac{\mathbf{r}}{r}, \tag{2.3}$$

where a dot indicates d/dr. Since the vectors $\mathbf{r}$, $\mathbf{r}'$, and $\mathbf{r}''$ are connected by a linear relation, the desired result follows. Actually, the same holds in a shell-type spherical medium, where the index function is discontinuous across the interfaces. The more general result holds because of the nature of Snell's law for the refraction of rays at the interfaces.

It is relatively simple to produce conventional glass lens elements containing spherical gradients. The procedure is to treat a spherical surface by the ion-diffusion method, thus creating a refractive index that varies with the distance along the inward normal to the surface. The index function then has spherical symmetry about the center of curvature of the surface. If desired, after creation of the gradient, the lens surface can be ground to a new shape, either nonspherical or spherical, but with a new center of curvature. On the other hand, gradients produced by ion diffusion do not extend very far into the glass, and it appears that spherical gradients are less useful in lens design than other types of gradients.

2.2   Determining the Rays

Since each ray in a spherical gradient lies in a plane through the origin, it is appropriate to adopt polar coordinates in the plane of a ray. With the help of Eq. (2.1) and the formula

$$ds = (1 + r^2\dot{\theta}^2)^{1/2}\,dr, \tag{2.4}$$

Eq. (1.2) takes the form

$$L = \int_{r_0}^{r} n(r)(1 + r^2\dot{\theta}^2)^{1/2}\,dr. \tag{2.5}$$

Here we regard the integrand as a function of r, θ, and $\dot{\theta}$, with r as independent variable. The corresponding Euler equation gives

$$\left(\frac{d}{dr}\right)[nr^2\dot{\theta}(1 + r^2\dot{\theta}^2)^{-1/2}] = 0. \tag{2.6}$$

This shows that the quantity e defined by

$$nr^2\dot{\theta}(1 + r^2\dot{\theta}^2)^{-1/2} = e \tag{2.7}$$

is invariant along any ray. However, e always has the same sign as $\dot{\theta}$ and so changes sign at points of the ray where $\dot{\theta}$ does.

The geometrical meaning of e is easily deduced from Fig. 2.1, which applies to any plane curve described in polar coordinates with $\theta' \geq 0$. With ψ defined as the angle between the tangent and the radius vector

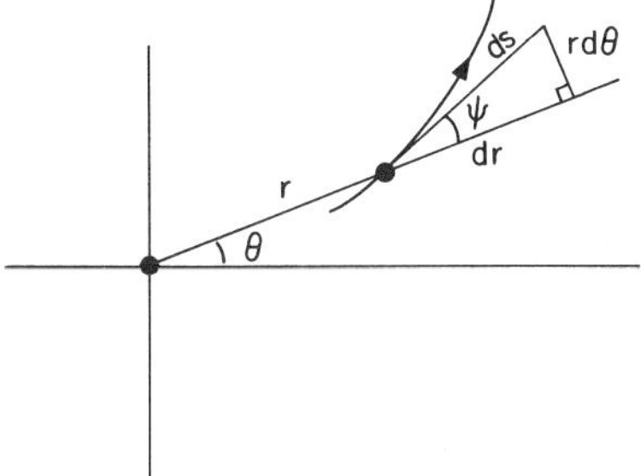

Fig. 2.1. Geometrical meaning of ψ.

from the origin, we have

$$\sin \psi = r\frac{d\theta}{ds} = r\frac{\dot{\theta}}{\dot{s}}$$

$$= r|\dot{\theta}|(1 + r^2\dot{\theta}^2)^{-1/2}, \qquad 0 \le \psi < \pi. \tag{2.8}$$

Hence,

$$e = \pm nr\sin\psi, \tag{2.9}$$

with $\pm$ selected as $\dot{\theta} \gtrless 0$. This relation is Snell's law generalized to a spherically symmetrical inhomogeneous medium.

The quantity $|e|$ is a scalar invariant along any ray in such a medium. There is, in fact, a vector invariant, namely,

$$\mathbf{p} = n(\mathbf{r} \times \mathbf{r}'). \tag{2.10}$$

For differentiation gives

$$\mathbf{p}' = n'(\mathbf{r} \times \mathbf{r}') + n(\mathbf{r} \times \mathbf{r}''), \tag{2.11}$$

after which elimination of $\mathbf{r}''$ by means of Eq. (2.3) shows that $\mathbf{p}' = 0$ at each point of the ray. As usual, exceptions can occur when the pertinent derivatives do not exist.

The polar equation of any ray can be written in terms of a single quadrature. The result, obtained by solving Eq. (2.7), is

$$\theta = \theta_0 + e \int_{r_0}^{r} \frac{dr}{r(n^2r^2 - e^2)^{1/2}}. \tag{2.12}$$

Here r_0 and θ_0 are values at any convenient initial point on the ray, and e can be found by evaluating Eq. (2.9) at this or any other point on the ray.

Although general and exact, Eq. (2.12) is not always convenient for use in practical problems when written in its present form. One drawback is that it involves polar coordinates in the plane of the ray rather than the coordinates of the entire optical system. Also the integral becomes improper at certain points of certain rays.

2.3 Maxwell's Fisheye Lens

As mentioned in Chapter 1, the fisheye lens of Maxwell represents an example of an absolute optical instrument in that it images sharply every point of a region of space. Its performance is based on the properties of a spherical gradient described by

$$n = \frac{n_0}{1 + (r/a)^2}, \tag{2.13}$$

where n_0 and a are constants. The action of this type of lens has been discussed by Herzberger (1958) and Born and Wolf (1970).

Each ray in a fisheye lens is a circular arc, which can be shown as follows. The slope angle τ of a ray is given by $\psi + \theta$, as seen from Fig. 2.1. Hence, the curvature K at any point can be written

$$K = \tau' = \psi' + \theta'. \tag{2.14}$$

By logarithmic differentiation of Eq. (2.9) we have

$$\frac{n'}{n} + \frac{r'}{r} + \frac{\psi'}{\tan \psi} = 0. \tag{2.15}$$

With the help of Fig. 2.1 it is seen that this can be written

$$\psi' = -\theta'\left[1 + \frac{rn'}{nr'}\right], \tag{2.16}$$

where again it is assumed that $\theta' \geq 0$. Thus

$$K = -\frac{rn'\theta'}{nr'} = -r\theta'\frac{\dot{n}}{n}. \tag{2.17}$$

From Eqs. (2.8) and (2.9) we have

$$\theta' = \frac{|e|}{nr^2}, \tag{2.18}$$

and from Eq. (2.13)

$$\dot{n}/n = -\frac{2nr}{n_0 a^2}.$$ (2.19)

It follows that

$$K = \frac{2|e|}{n_0 a^2}.$$ (2.20)

Since the curvature is constant, the ray lies in a circular arc.

To verify the stigmatic imaging we picture the spherical medium with center of symmetry at the origin O, as shown in Fig. 2.2. We first consider a sphere about O of radius a, where a is the constant appearing in Eq. (2.13), and picture a ray intersecting the sphere $r = a$ at a point A, as shown in the figure.

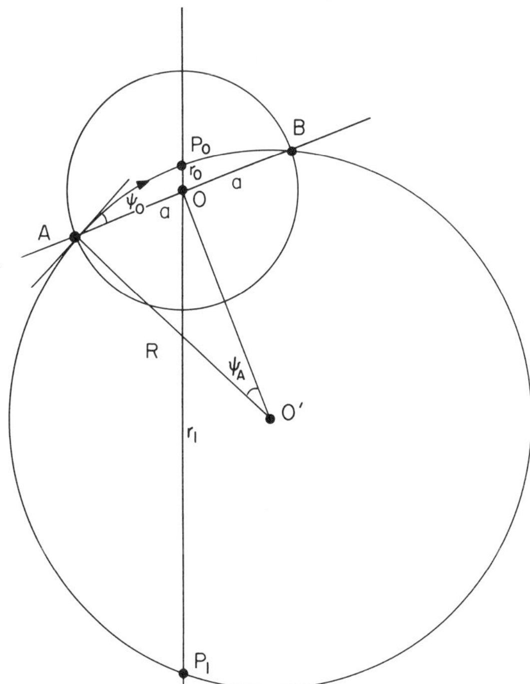

Fig. 2.2. Geometry of the fisheye lens.

The ray, following a circular arc centered at a point O', will intersect the sphere $r = a$ again at some point B. We shall see that the chord AB has length $2a$, so that A and B are on opposite ends of a diameter of the sphere $r = a$. To see this we note, from Eq. (2.13), that the index at A and B is $n_0/2$, so that from Eq. (2.9)

$$e = \frac{an_0}{2} \sin \psi_A, \tag{2.21}$$

for the case with $\dot{\theta} > 0$. Inspection of Fig. (2.2) gives

$$\frac{AB}{2} = R \sin \psi_A = \frac{\sin \psi_A}{K}. \tag{2.22}$$

From this and Eqs. (2.20) and (2.21), it follows that $AB = 2a$.

Since every ray leaving a point A on the sphere $r = a$ must intersect the sphere again on the opposite end B of a diameter, it follows that every such point A is sharply imaged. It remains to show that a general point not on this sphere is also sharply imaged.

We consider a general point P_0 at a distance r_0 from O, as indicated in Fig. 2.2. Consider also a general ray lying on a circular arc passing through P_0 and intersecting the sphere $r = a$ at two points A and B. As shown above, A and B must lie on opposite ends of a diameter of this sphere. Let r_1 be the distance OP_1 at which the ray again intersects the line OP_0. From simple geometry

$$r_0 r_1 = a^2, \tag{2.23}$$

showing that r_1 is the same for all rays passing through P_0. Hence P_0 is sharply imaged at P_1. It is interesting to note from Eq. (2.23) that the imaging process is simply that of inversion with respect to the sphere $r = a$.

2.4   The Luneburg Lens

The Maxwell fisheye lens has the drawback that only points within or on the surface of the lens are sharply imaged. A lens proposed by Luneburg images sharply every parallel bundle of rays incident on the outer surface of the lens. Here again, the index function has the form of a spherical gradient, which in this case is given by

$$n = \left[2 - \left(\frac{r}{r_0} \right)^2 \right]^{1/2}. \tag{2.24}$$

It is sufficient to analyze a typical ray lying in the meridional plane, as shown in Fig. 2.3. The ray is assumed to be incident horizontally at a point P_0. As Eq. (2.24) shows, the index value of the lens at P_0 is unity, matching that of the surrounding medium (assumed to be air). Consequently, the ray is not bent at P_0 and remains horizontal just inside the lens.

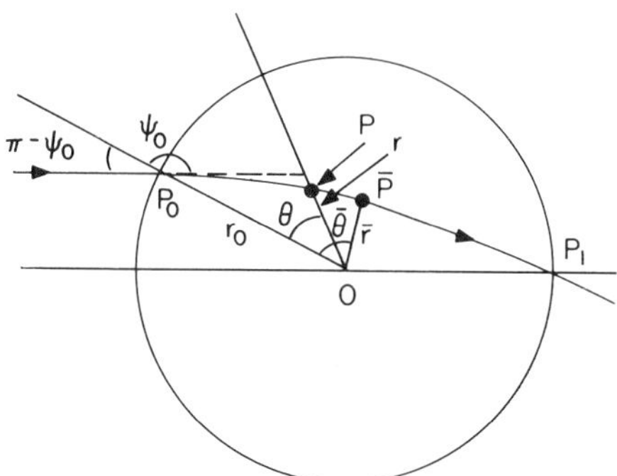

Fig. 2.3.   Geometry of the Luneburg lens.

The polar equation of the ray is found from Eq. (2.12) by substituting for n the function given in Eq. (2.24), the result being

$$\theta = \theta_0 + e \int_{r_0}^{r} \frac{dr}{r[r^2[2 - (r/r_0)^2] - e^2]^{1/2}}. \tag{2.25}$$

Here e is given by Eq. (2.9) as

$$e = -r_0 \sin \psi_0, \tag{2.26}$$

with $\dot{\theta} < 0$ at P_0.

Changing the variable of integration by the relation

$$v = \left(\frac{r_0}{r}\right)^2 \tag{2.27}$$

reduces Eq. (2.25) to the form

$$\theta = \theta_0 + \frac{1}{2}\sin \psi_0 \int_1^v \frac{dv}{(2v - 1 - v^2 \sin^2 \psi_0)^{1/2}}. \tag{2.28}$$

The original integral has a singularity at the point $\bar{P}$ at which $d\theta/dr$ becomes infinite. The corresponding value of v is found by setting

$$2v - 1 - v^2 \sin^2 \psi_0 = 0, \tag{2.29}$$

giving

$$\frac{1}{\bar{v}} = \left(\frac{\bar{r}}{r_0}\right)^2 = 1 \pm \cos \psi_0. \tag{2.30}$$

In order to have $\bar{r}/r_0 < 1$, it is necessary to select the negative sign, so that

$$\frac{\bar{r}}{r_0} = \sqrt{2} \sin\left(\frac{\psi_0}{2}\right). \tag{2.31}$$

It is clear from the symmetry of the index function and the principle of reversibility of light rays that $\bar{P}$ must lie midway between the points P_0 and P_1 where the ray meets the circumference. For the ray from P_0 to P_1 must follow the same curve as a ray from P_1 to P_0.

With the reference line for angle θ chosen as OP_0, the integration in Eq. (2.28) taken from P_0 to $\bar{P}$ gives

$$\bar{\theta} = \frac{1}{2} \sin \psi_0 \int_1^v \frac{dv}{(2v - 1 - v^2 \sin^2 \psi_0)^{1/2}}. \tag{2.32}$$

This leads to

$$2\bar{\theta} = \sin^{-1}\left[\frac{1 - v \sin^2 \psi_0}{\cos \psi_0}\right]_1^v$$

$$= \psi_0 - \pi. \tag{2.33}$$

From Fig. 2.3 it is then clear that the second point P_1 where the ray intersects the sphere $r = r_0$ is such that OP_1 is parallel to the original ray. Hence all rays of the bundle pass through the same point P_1.

Following Luneburg, whose lens was announced about 1944, several investigations were made into the properties of spherical gradient lenses (see Eaton, 1953; Stettler, 1955; Toraldo di Francia, 1957; Huynen, 1958; Uslenghi, 1964). In most of these studies it was assumed that the refractive index at the outer surface was unity. Such media can hardly be made for use with visible light but might be useful for microwave applications.

2.5 The Generalized Luneburg Lens

The lens introduced by Luneburg is restricted in that only object points at infinity are sharply imaged. Morgan (1958) extended Luneburg's idea and demonstrated the possibility of a spherical gradient lens that images one finite sphere sharply onto another. In some cases both the object and image surfaces lie outside the lens itself.

In studying this problem it is sufficient to consider only rays lying in a single plane through the origin (point of symmetry) as indicated in Fig. 2.4. Also, there is no loss of generality in assuming the radius of the lens to be unity. Following Morgan we set

$$\rho = nr \tag{2.34}$$

as a convenient parameter. Evidently specifying either ρ or n as a function of r determines the other function.

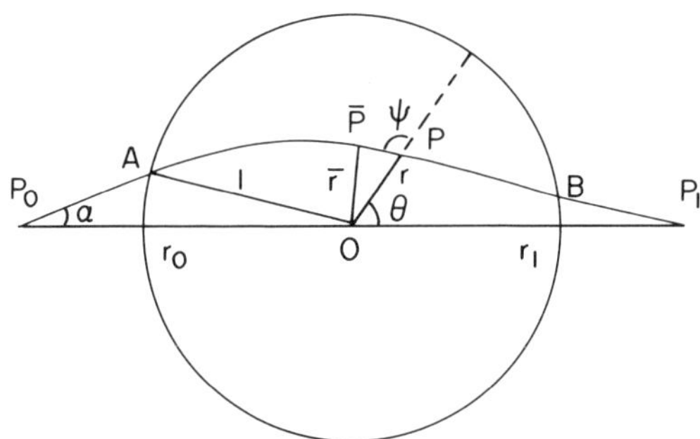

Fig. 2.4. Generalized Luneburg lens.

It is helpful to assume that $\rho(r)$ is merely a piecewise continuous function (admitting a finite number of discontinuities). This device allows us to consider the exterior of the lens to be part of an enlarged medium that can be described by the single function $\rho(r)$. If $n = 1$ in the region $r > 1$, we have $\rho = r$ there. For $r > 1$, we shall assume $\rho(r)$ to be strictly positive, except that $\rho(0) = 0$.

In view of Snell's law the quantity k defined by

$$k = |e| = nr \sin \psi \tag{2.35}$$

is invariant along a ray not only inside and outside the lens but also in passage across the surface $r = 1$. From Fig. 2.4 we have

$$k = r_0 \sin \alpha. \tag{2.36}$$

In applying Eq. (2.12) to a ray from P_0 to P_1 it must be recalled that e changes sign when $\dot{\theta}$ does. Accordingly, the integration must be divided into two parts corresponding to the intervals $P_0\bar{P}$ and $\bar{P}P_1$, $\bar{P}$ being the point where $dr/d\theta = 0$. In the present treatment we assume there is only one such point. Taking OP_1 as the reference line for angle θ, we have, therefore,

$$0 = \pi + \int_{r_0}^{\bar{r}} \frac{k\,dr}{r(\rho^2 - k^2)^{1/2}} - \int_{\bar{r}}^{r_1} \frac{k\,dr}{r(\rho^2 - k^2)^{1/2}}. \tag{2.37}$$

This equation can be arranged in the form

$$\int_{\bar{r}}^{1} \frac{k\,dr}{r(\rho^2 - k^2)^{1/2}} = f(k), \qquad 0 \le k \le 1, \tag{2.38}$$

where

$$f(k) = \frac{1}{2} \int_{r_0}^{1} \frac{k\,dr}{r(r^2 - k^2)^{1/2}} + \frac{1}{2} \int_{r_1}^{1} \frac{k\,dr}{r(r^2 - k^2)^{1/2}} + \frac{\pi}{2}$$

$$= \frac{1}{2} \sin^{-1}\left(\frac{k}{r_0}\right) + \frac{1}{2} \sin^{-1}\left(\frac{k}{r_1}\right) + \cos^{-1} k. \tag{2.39}$$

With $\bar{r}$ regarded as a function of k, Eq. (2.38) represents an integral equation for the determination of a function $\rho(r)$ for which P_0 is imaged sharply at P_1.

Figure 2.4 and Eqs. (2.38) and (2.39) apply for the case with two external foci ($r_0 > 1$, $r_1 > 1$). With slight modifications one can treat the case with one internal and one external focus (see Morgan, 1958). Luneburg's original lens is a limiting case in which P_0 is at infinity and P_1 is on the surface of the lens.

Before solving Eq. (2.38) we note that, if the index of the lens at $r = 1$ is less than unity, some rays from P_0 will be totally reflected. Then the full aperture of the lens will not be used.

Let us introduce a value a such that

$$\rho(a) = 1 \tag{2.40}$$

and rewrite Eq. (2.38) in the form

$$\int_{\bar{r}}^{a} \frac{k\,dr}{r(\rho^2 - k^2)^{1/2}} = f(k) - F(k), \tag{2.41}$$

where

$$F(k) = \int_{a}^{1} \frac{k\,dr}{r(\rho^2 - k^2)^{1/2}}. \tag{2.42}$$

Then, if $\rho(r)$ is a continuous, strictly increasing function of r in $0 < r < a$, likewise $r(\rho)$ is a continuous, strictly increasing function of ρ in the interval $0 \le \rho \le 1$. We can then introduce a new function $g(\rho)$ by

$$\log r = -g(\rho),$$
$$\frac{dr}{r} = -g'(\rho)\,d\rho, \tag{2.43}$$

so that Eq. (2.41) reduces to

$$\int_{1}^{k} \frac{kg'(\rho)\,d\rho}{(\rho^2 - k^2)^{1/2}} = f(k) - F(k) \tag{2.44}$$

since $\rho = k$ at $r = \bar{r}$.

With a further change of variable of integration this equation can be put in the form of the well-known Abel integral equation, for which the solution is known. However, we solve the equation directly. After replacing the dummy variable ρ by σ, multiplying each side by $dk/(k^2 - \rho^2)^{1/2}$, and integrating from ρ to 1, we have

$$\int_{\rho}^{1} \int_{1}^{k} \frac{kg'(\sigma)\,d\sigma\,dk}{[(\sigma^2 - k^2)(k^2 - \rho^2)]^{1/2}} = \int_{\rho}^{1} \frac{[f(k) - F(k)]\,dk}{(k^2 - \rho^2)^{1/2}}. \tag{2.45}$$

Changing the order of integration gives

$$\int_{1}^{\rho} \left[\int_{\rho}^{\sigma} \frac{k\,dk}{(\sigma^2 - k^2)^{1/2}(k^2 - \rho^2)^{1/2}} \right] g'(\sigma)\,d\sigma$$
$$= \frac{1}{2}\pi[g(\rho) - g(1)] = \int_{\rho}^{1} \frac{[f(k) - F(k)]\,dk}{(k^2 - \rho^2)^{1/2}}. \tag{2.46}$$

It follows that

$$\log\left(\frac{a}{r}\right) = \frac{2}{\pi} \int_{\rho}^{1} \frac{[f(k) - F(k)]\,dk}{(k^2 - \rho^2)^{1/2}}, \qquad 0 \le \rho \le 1. \tag{2.47}$$

As suggested originally by Luneburg, we now introduce a function $\omega(\rho, s)$ defined by

$$\omega(\rho, s) = \frac{1}{\pi} \int_{\rho}^{1} \frac{\sin^{-1}(k/s)\,dk}{(k^2 - \rho^2)^{1/2}}, \qquad 0 \le \rho \le 1, \quad s \ge 1. \tag{2.48}$$

Numerical values of this function are tabulated by Morgan. In particular it is noted that

$$\omega(0, s) = \frac{1}{\pi} \int_{0}^{1/s} \sin^{-1} x \frac{dx}{x},$$

$$\omega(\rho, 1) = \frac{1}{2} \log[1 + (1 - \rho^2)^{1/2}], \tag{2.49}$$

$$\omega(1, s) = 0, \qquad \omega(\rho, \infty) = 0.$$

The second of these formulas was proved by Luneburg.

With the above definition of ω, Eq. (2.39) leads to

$$\frac{2}{\pi} \int_{\rho}^{1} \frac{f(k)\,dk}{(k^2 - \rho^2)^{1/2}} = \omega(\rho, r_0) + \omega(\rho, r_1) - \log\rho. \tag{2.50}$$

Here. the value of the last term can be obtained by replacing $\cos^{-1} k$ by $(\pi/2) - \sin^{-1} k$ in Eq. (2.39) and taking advantage of Eq. (2.49).

In considering $F(k)$, as defined by Eq. (2.42), it is helpful to define a function $P(r)$ such that

$$F(k) = \int_{a}^{\bar{r}} \frac{k\,dr}{r(\rho^2 - k^2)^{1/2}} = \int_{a}^{1} \frac{k\,dr}{r[P^2(r) - k^2]^{1/2}}, \qquad 0 \le k \le 1. \tag{2.51}$$

In practice, $P(r)$ must usually be evaluated by numerical methods. However, assuming that such a function has been determined, we define a function $\Omega(\rho)$ by

$$\Omega(\rho) = \frac{2}{\pi} \int_{\rho}^{1} \frac{F(k)\,dk}{(k^2 - \rho^2)^{1/2}}, \tag{2.52}$$

which reduces to

$$\Omega(\rho) = \frac{2}{\pi} \int_a^1 \tan^{-1} \left[\frac{1 - \rho^2}{P^2(r) - 1} \right]^{1/2} \frac{dr}{r}. \tag{2.53}$$

Substitution of these relations into Eq. (2.47) and taking note of Eq. (2.50) gives

$$\log\left(\frac{a}{r}\right) = \omega(\rho, r_0) + \omega(\rho, r_1) - \log \rho - \Omega(\rho), \tag{2.54}$$

from which we find the result

$$n = \frac{1}{a} \exp[\omega(\rho, r_0) + \omega(\rho, r_1) - \Omega(\rho)],$$

$$\tag{2.55}$$

$$n = \frac{\rho}{r}.$$

These equations determine the refractive index n as a function of r, but only by a parametric representation, the parameter ρ running from 0 to 1.

If $a = 1$, we have $\Omega(\rho) = 0$, leading to Luneburg's solution. For we now have

$$n = \exp[\omega(\rho, r_0) + \omega(\rho, r_1)],$$

$$\tag{2.56}$$

$$n = \frac{\rho}{r}.$$

Taking $r_0 = \infty$ and $r_1 = 1$, we find, with the help of Eq. (2.49),

$$n^2 = 1 + \left[1 - (nr)^2 \right]^{1/2}, \tag{2.57}$$

which reduces to

$$n = (2 - r^2)^{1/2}, \tag{2.58}$$

in agreement with Eq. (2.24).

In order to obtain solutions for other positions of P_0 and P_1, the procedure is to make appropriate selections of the constant a and the function $P(r)$. Morgan has established the existence of solutions on the condition that

$$\sin^{-1}\left(\frac{1}{r_0}\right) + \sin^{-1}\left(\frac{1}{r_1}\right) \geq 2 \int_a^1 \frac{dr}{r[P^2(r) - 1]^{1/2}}. \qquad (2.59)$$

It is usually not feasible to make lenses of these types for use in visible light. For example, in Eq. (2.58), the refractive index must vary from $\sqrt{2}$ at the center to unity at the surface of the lens. However, for microwave applications, the required refractive index distributions can often be produced approximately by various devices, and the above theory has proved useful in designing wide-angle radar scanners and antennas.

The theory of generalized Luneburg lenses has found application in the field of integrated optics. In planar wave guides, a gradient index medium can be simulated by varying the thickness of the planar layers or by creating depressions in the layers. Southwell (1977a,b) describes mathematical methods for determining the effective index profiles required for two-dimensional imaging in this application. His papers contain helpful suggestions for the numerical solution of the integral equation (Eq. (2.38)) arising in the theory of generalized Luneburg lenses.

Another application of the theory given in the preceding sections has arisen recently in the problem of measuring the index profiles in radial gradient fibers. If a beam is sent through a fiber in a direction perpendicular to its axis, the rays obey the laws for a spherical gradient. The determination of the index function depends on the solution of an integral equation like that in Eq. (2.38). Furthermore, by using an immersion liquid, refraction at the fiber surface can be avoided.

2.6 Astronomical Refraction

Early investigations of astronomical refraction were based on the assumption that the atmosphere is a homogeneous spherical shell. Later the theory was refined by approximating the atmosphere by a system of concentric shells. The mathematical difficulties associated with this approach are considerable.

Recently White (1975) proposed a simple analytical model, where a spherical gradient given by

$$(nr)^k \log n = a = (n_0 r_0)^k \log n_0 \qquad (2.60)$$

is assumed, k and a being constants. Here the pertinent integrals can be evaluated in closed form, and the function given by the above equation appears to fit measured index data reasonably well. For example, with $k > 1$, we see that

$$\lim_{r \to \infty} n = 1, \tag{2.61}$$

whereas with a previous model given by

$$nr^k = a, \tag{2.62}$$

with $k > 0$, attributed to Simpson,

$$\lim_{r \to \infty} n = 0 \tag{2.63}$$

For that reason, in using Simpson's formula it has been customary to assume $n = 1$ for r above some fixed value.

The crucial integrals required for the astronomical refraction problem are for the polar angle θ_{12} between two points and the ray deviation δ_{12} between the two points. From Fig. 2.1 we have

$$\tan \psi = \frac{r \, d\theta}{dr} \tag{2.64}$$

giving

$$\theta_{12} = \int_{\log r_1}^{\log r_2} \tan \psi \, d \log r. \tag{2.65}$$

With a suitable sign convention, so that

$$\theta_{12} + \delta_{12} = \psi_2 - \psi_1, \tag{2.66}$$

White obtains the formula

$$\delta_{12} = \int_{\log n_1}^{\log n_2} \tan \psi \, d \log n. \tag{2.67}$$

Equation (2.9) gives

$$\tan \psi = e(n^2 r^2 - e^2)^{-1/2}, \tag{2.68}$$

where

$$e = \pm nr \sin \psi = n_0 r_0 \sin \psi_0. \tag{2.69}$$

Thus

$$\theta_{12} = e \int_{\log r_1}^{\log r_2} (n^2 r^2 - e^2)^{-1/2}\, d\log r \tag{2.70}$$

and

$$\delta_{12} = e \int_{\log n_1}^{\log n_2} (n^2 r^2 - e^2)^{-1/2}\, d\log n. \tag{2.71}$$

Only one of these integrals needs to be evaluated since the values of ψ_1 and ψ_2 are found from Eq. (2.69), and the value of the other integral can be obtained by Eq. (2.66). Assuming Eq. (2.60) and

$$v = \log n, \tag{2.72}$$

we have, by noting Eq. (2.69),

$$v = a(nr)^{-k} = a\left(\frac{\sin\psi}{|e|}\right)^k,$$

$$dv = ak\left(\frac{\sin\psi}{|e|}\right)^k\left(\frac{d\psi}{\tan\psi}\right). \tag{2.73}$$

Hence, from Eq. (2.67)

$$\delta_{12} = ak|e|^{-k} \int_{\psi_1}^{\psi_2} \sin^k \psi\, d\psi. \tag{2.74}$$

Using Eqs. (2.60) and (2.69), we then obtain the formula

$$\delta_{12} = k(\log n_0)\sin^{-k}\psi_0 \int_{\psi_1}^{\psi_2} \sin^k \psi\, d\psi. \tag{2.75}$$

If k is an even integer, the integral can be evaluated in closed form in terms of elementary functions by using well-known reduction formulas. There is evidence to indicate that $k = 900$ is a reasonable empirical value.

It is found from this analysis that the model suggested by White is not only more satisfactory from the physical viewpoint than that of Simpson, but also leads to relatively simple evaluation of the integrals involved. White gives results of computations of astronomical refraction and shows that his approach gives better agreement with measured results than those obtained by earlier models. In his paper he indicates a number of other functions for which either Eq. (2.70) or (2.71) can be evaluated in closed form.

Chapter 3
RAY TRACE IN A
SPHERICAL GRADIENT

3.1 Basic Equations

If a spherical gradient is to be used in an optical system having a construction generally similar to conventional systems, it is advisable to revise the ray equations presented in Chapter 2. For one thing, the equations given there are expressed in a local coordinate system and so should be converted to the global coordinates, which apply to the entire optical system. Furthermore, the equations as written in the last chapter become singular for certain rays. Also, in many cases the center of symmetry is situated far from the region of interest, so that some coordinates become excessively large or even infinite. More practical formulas are derived in this chapter following Marchand (1970).

In a spherical medium (n a function of the distance r from an origin at O), it is natural to adopt spherical coordinates (r, θ, ϕ). However, if the z-axis is the axis of the optical system, it is convenient to define θ and ϕ not relative to this axis, but relative to a line OP_0 from the origin to a starting point P_0 on the ray (see Fig. 3.1).

As shown in the figure, α_0 is the angle between OP_0 and the z-axis, and P, with coordinates r, θ, and ϕ, is a typical point on the ray (ϕ not being

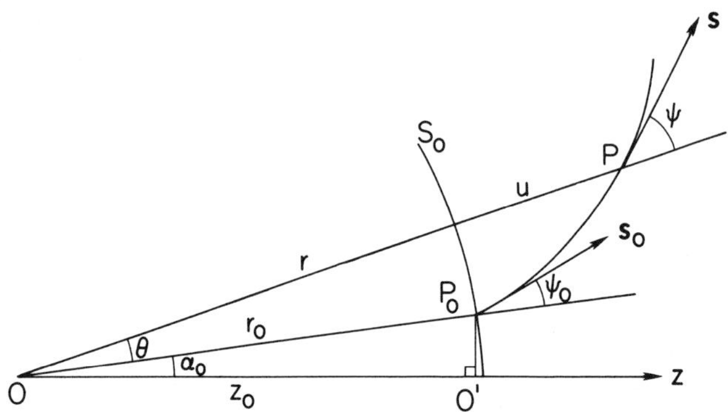

Fig. 3.1.   Spherical gradient, $z_0 > 0$, $\varepsilon = +1$ (from Marchand, 1970).

shown in the figure). Figure 3.1 is schematic in that the plane OP_0P need not coincide with the plane OP_0z.

The quantity u in the figure is simply the penetration distance $r - r_0$, and ψ is the angle between the radial and tangential vectors. The vectors $\mathbf{s}_0$ and $\mathbf{s}$ are the optical tangential vectors of the ray at P_0 and P, having lengths n_0 and n respectively.

Figure 3.1 applies when P_0 lies to the right ($z_0 > 0$) of the center of symmetry O of the index function. Figure 3.2 shows the corresponding case when P_0 lies to the left of O ($z_0 < 0$). In the latter case it is convenient to define u as $r_0 - r$, so that in both cases u represents the penetration depth of a point P beyond the sphere S_0 of constant index passing through P_0. For most rays of practical interest u will remain nonnegative in both cases.

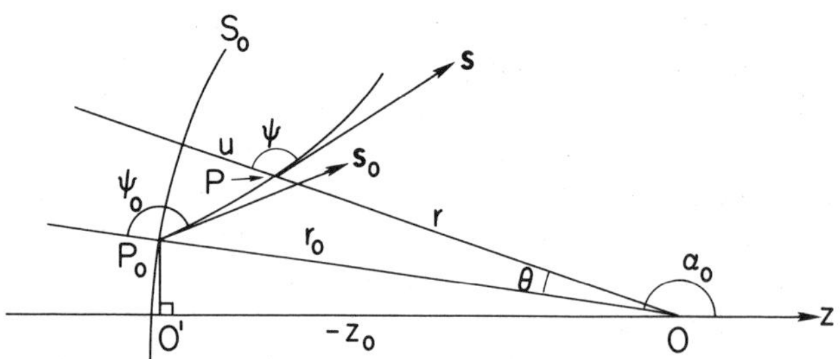

Fig. 3.2.   Spherical gradient, $z_0 < 0$, $\varepsilon = -1$ (from Marchand, (1970).

For the local coordinate system chosen the light-path integral has the form

$$L = \int_{s_0}^{s} n \, ds = \int_{r_0}^{r} F \, dr \tag{3.1}$$

with

$$F = n(r)(1 + r^2\dot{\theta}^2 + r^2 \sin^2\theta \dot{\phi}^2)^{1/2}, \tag{3.2}$$

the dot indicating d/dr. Since $\partial F/\partial\phi = 0$, the first Euler equation reduces to

$$nr^2 \sin^2\theta \, \dot{\phi}(1 + r^2\dot{\theta}^2 + r^2 \sin^2\theta \, \dot{\phi}^2)^{-1/2} = \text{const.} \tag{3.3}$$

For a nonradial ray $\sin\theta \neq 0$ except at P_0, where $\theta = 0$. It follows that $\dot{\phi} = 0$ at all points, and ϕ is constant along the ray. This confirms the fact previously shown that every ray in a spherical medium lies along a plane curve.

The second Euler equation reduces to

$$nr^2\dot{\theta}(1 + r^2\dot{\theta}^2)^{-1/2} = e, \tag{3.4}$$

where

$$e = \pm nr \sin\psi, \tag{3.5}$$

e being constant along the ray except for changes of sign. From Eq. (3.4) it is clear that e always has the same sign as $\dot{\theta}$, depending on whether r is an increasing or decreasing function of θ. The derivation of Eq. (3.5) from (3.4) was discussed in the last chapter.

Equations (3.4) and (3.5) agree with Eqs. (2.7) and (2.9), which were derived in a slightly different manner. The geometrical meaning of ψ is clear from Figs. 3.1 and 3.2. The value of e can be obtained from Eq. (3.5) applied to the point P_0, i.e.,

$$e = \pm n_0 r_0 \sin\psi_0. \tag{3.6}$$

For the analysis of most conventional types of optical systems it is helpful to assume $r \geq 0$, $0 \leq \theta < \pi/2$, and $0 \leq \psi < \pi$, throughout. Also we consider two cases according to whether $z_0 > 0$ (Fig. 3.1 with $\dot{\theta} \geq 0$, $e \geq 0$, and $\psi \leq \pi/2$) or $z_0 < 0$ (Fig. 3.2 with $\dot{\theta} < 0$, $e < 0$, and $\psi > \pi/2$).

In order to include both cases in one analysis we define the symbol

$$\varepsilon = \operatorname{sgn} z_0 = \pm 1 \quad \text{as} \quad z_0 \gtrless 0. \tag{3.7}$$

Equation (3.4) can be solved for either case in terms of a quadrature:

$$\theta = e \int_{r_0}^{r} \frac{dr}{r(n^2 r^2 - e^2)^{1/2}} \tag{3.8}$$

with n a function of r and

$$e = \varepsilon n r \sin \psi = \varepsilon n_0 r_0 \sin \psi_0. \tag{3.9}$$

The result, however, refers to a tilted axis lying along OP_0. We proceed to convert to the global axial system.

Figure 3.3 is similar to Fig. 3.1 but shows radial and tangential vectors as well as the corresponding unit vectors to be used in the analysis. The vectors, defined relative to the untilted Cartesian coordinate system, are defined by

$$
\begin{aligned}
\mathbf{r} &= (x, y, z), & \mathbf{r}^* &= \frac{\mathbf{r}}{r}, \\[2ex]
\mathbf{r}_0 &= (x_0, y_0, z_0), & \mathbf{r}_0^* &= \frac{\mathbf{r}_0}{r_0}, \\[2ex]
\mathbf{s} &= (p, q, l), & \mathbf{s}^* &= \frac{\mathbf{s}}{n}, \\[2ex]
\mathbf{s}_0 &= (p_0, q_0, l_0), & \mathbf{s}_0^* &= \frac{\mathbf{s}_0}{n_0}.
\end{aligned}
\tag{3.10}
$$

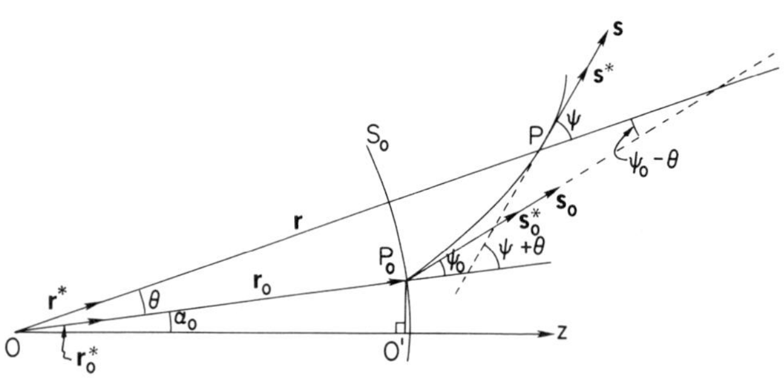

Fig. 3.3. Spherical gradient and related vectors (from Marchand, 1970).

Since the ray lies in a plane through the origin, we have

$$\mathbf{r}^* = \alpha \mathbf{r}_0^* + \beta \mathbf{s}_0^*, \tag{3.11}$$

where α and β are coefficients to be determined. By scalar multiplication we find

$$\begin{aligned}
\mathbf{r}^* \cdot \mathbf{r}_0^* &= \alpha + \beta(\mathbf{s}_0^* \cdot \mathbf{r}_0^*), \\
\mathbf{r}^* \cdot \mathbf{s}_0^* &= \alpha(\mathbf{r}_0^* \cdot \mathbf{s}_0^*) + \beta,
\end{aligned} \tag{3.12}$$

showing that

$$\begin{aligned}
\cos \theta &= \alpha + \beta \cos \psi_0, \\
\cos(\psi_0 - \theta) &= \alpha \cos \psi_0 + \beta.
\end{aligned} \tag{3.13}$$

It follows that

$$\begin{aligned}
\beta &= \sin \theta / \sin \psi_0, \\
\alpha &= \cos \theta - \beta \cos \psi_0.
\end{aligned} \tag{3.14}$$

Writing Eq. (3.11) in component form gives

$$\begin{aligned}
x &= r\left(\alpha \frac{x_0}{r_0} + \beta \frac{p_0}{n_0} \right), \\
y &= r\left(\alpha \frac{y_0}{r_0} + \beta \frac{q_0}{n_0} \right), \\
z &= r\left(\alpha \frac{z_0}{r_0} + \beta \frac{l_0}{n_0} \right).
\end{aligned} \tag{3.15}$$

Since α and β are given in terms of θ by Eqs. (3.14) and θ in terms of r by Eq. (3.8), Eqs. (3.15) represent the essential ray-tracing equations relative to the untilted system. They are, however, still not in the best form for use in practical problems.

3.2 Improving the Ray-Trace Formulas

Equations (3.15), in conjunction with Eqs. (3.14) and (3.8), give x, y, and z in terms of the parameter r when the appropriate eliminations are carried out. However, with the origin chosen at the point of symmetry of the gradient, both r and z can be excessively large. Another difficulty with these equations is that β becomes indeterminate for rays that are nearly radial (θ small) in the spherical medium. These defects can be removed by suitable manipulation of the equations.

We first define a quantity M_0 by

$$M_0 = \int_{r_0}^{r} \frac{dr}{r(n^2r^2 - e^2)^{1/2}}, \tag{3.16}$$

so that

$$\theta = eM_0. \tag{3.17}$$

Then, from Eq. (3.9) and the first of Eqs. (3.14), we have

$$\beta = \varepsilon n_0 r_0 M_0 \operatorname{sinc} \theta, \tag{3.18}$$

where

$$\operatorname{sinc} \theta = \frac{\sin \theta}{\theta}. \tag{3.19}$$

The numerical evaluation of $\operatorname{sinc} \theta$ for small θ values is easily performed by means of a McLaurin's series:

$$\operatorname{sinc} \theta = 1 - \frac{\theta^2}{3!} + \frac{\theta^4}{5!} - \cdots. \tag{3.20}$$

As defined above

$$u = |r - r_0| = \varepsilon(r - r_0) \tag{3.21}$$

for the two cases with $z_0 \gtrless 0$. It is now helpful to define ρ_0 as the curvature of the sphere S_0 of constant index passing through P_0. With r_0 considered always positive, we assume ρ_0 to be positive or negative as the sphere S_0 is concave to the right or left. This agrees with the customary convention on the curvature of optical surfaces. It follows from Eq. (3.21) that

$$\frac{r}{r_0} = 1 - \rho_0 u \tag{3.22}$$

in both cases. In most practical applications both $|\rho_0|$ and u will be small quantities.

With the help of the definitions adopted for u and ρ_0 we now introduce u as a new variable of integration in Eq. (3.16). The result is

$$M_0 = \varepsilon M \rho_0^2, \tag{3.23}$$

where

$$M = \int_0^u \frac{du}{(1 - \rho_0 u)[n^2(1 - \rho_0 u)^2 - (e\rho_0)^2]^{1/2}}. \tag{3.24}$$

It is understood that, in order to use this formula, it is necessary to express the refractive index n as a function of u instead of r. The above integral for M is well behaved when OP_0 is large since $|\rho_0 u|$ and $|e\rho_0|$ do not assume excessively large values. In fact, it is convenient to introduce a scaled value defined by

$$\bar{e} = |e\rho_0| = n_0 \sin \psi_0. \tag{3.25}$$

Collecting the preceding results we have

$$\begin{aligned}
\theta &= \bar{e}|\rho_0|M, \qquad \beta = n_0|\rho_0|M \operatorname{sinc} \theta, \\
\alpha &= \cos \theta - \beta \cos \psi_0,
\end{aligned} \tag{3.26}$$

and

$$\begin{aligned}
x &= (1 - \rho_0 u)(\alpha x_0 + p_0 M \operatorname{sinc} \theta), \\
y &= (1 - \rho_0 u)(\alpha y_0 + q_0 M \operatorname{sinc} \theta), \\
z &= (1 - \rho_0 u)(\alpha z_0 + l_0 M \operatorname{sinc} \theta).
\end{aligned} \tag{3.27}$$

These equations, together with Eqs. (3.24) and (3.26), provide an improved set of ray-tracing formulas. Here, with the elimination of α, β, and θ, we have x, y, and z expressed in terms of the parameter u. However, with the origin at the center of symmetry O, z and z_0 can be very large or infinite. This difficulty can be removed by introducing as a new origin the point O', the projection of P_0 on the optical axis. Thus we define

$$\bar{z} = z - z_0. \tag{3.28}$$

With the above shift in origin, the third of Eqs. (3.27) can be written, after some manipulations,

$$u = \varepsilon \sec \theta \{\bar{z}/a + M[(1 - \rho_0 u)(b - l_0/a)\operatorname{sinc}\theta + \bar{e}\sin(\theta/2)\operatorname{sinc}(\theta/2)]\}, \tag{3.29}$$

where a and b are defined by

$$\begin{aligned} a &= \varepsilon[1 - \rho_0^2(x_0^2 + y_0^2)]^{1/2} = \cos\alpha_0 \\ b &= al_0 + |\rho_0|(\rho_0 x_0 + q_0 y_0) = n_0 \cos\psi_0. \end{aligned} \tag{3.30}$$

In principle, Eq. (3.29) can be solved for u in terms of $\bar{z}$ after eliminating θ and M by means of Eqs. (3.24) and (3.26). The result can be inserted in the first two of Eqs. (3.27), the overall effect being that x and y are determined in terms of $\bar{z}$. However, the elimination of θ and M from Eq. (3.29) leaves a transcendental equation that cannot be solved explicitly for u in terms of $\bar{z}$. Consequently, it is necessary to resort to numerical methods. We outline a routine for doing this.

With $\bar{z}$ given, a trial value of u (say 0) is selected and substituted on the right-hand side of Eq. (3.29). This suggests a new trial value of u. The procedure is repeated until the difference of successive values of u is sufficiently small. In the course of the iteration it is advisable to retain successive values of various quantities, as their terminal values are useful later. With such a computing routine the path of a ray is readily determined with good accuracy.

3.3　Finding the Direction of the Ray

In order to determine the direction of the ray at each point we recall that

$$p = nx' = n\frac{\dot{x}}{\dot{s}},$$

$$q = ny' = n\frac{\dot{y}}{\dot{s}}, \tag{3.31}$$

$$l = nz' = n\frac{\dot{z}}{\dot{s}}.$$

However, with the help of Eqs. (3.9) we have

$$1/\dot{s} = dr/ds = \cos\psi$$

$$= \varepsilon\left[1 - \frac{e^2}{n^2 r^2}\right]^{1/2}$$

$$= \varepsilon\left[1 - \frac{e^2 \rho_0^2}{n^2(1 - \rho_0 u)^2}\right]^{1/2} = \frac{t/n}{1 - \rho_0 u}, \tag{3.32}$$

where

$$t = \varepsilon\left[n^2(1 - \rho_0 u)^2 - \bar{e}^2\right]^{1/2}, \tag{3.33}$$

and $\bar{e}$ is given in Eq. (3.25). Hence

$$p = \frac{\dot{x}t}{1 - \rho_0 u} \tag{3.34}$$

with similar formulas for q and l.

Differentiation of the first of Eqs. (3.15) gives

$$\dot{x} = (\alpha + r\dot{\alpha})\frac{x_0}{r_0} + (\beta + r\dot{\beta})\frac{p_0}{n_0}. \tag{3.35}$$

But Eqs. (3.14) can be written

$$\alpha = \frac{\sin(\psi_0 - \theta)}{\sin\psi_0}$$

$$\beta = \frac{\sin\theta}{\sin\psi_0}, \tag{3.36}$$

so that

$$\dot{\alpha} = -\frac{\dot{\theta}\cos(\psi_0 - \theta)}{\sin\psi_0},$$

$$\dot{\beta} = \frac{\dot{\theta}\cos\theta}{\sin\psi_0}, \tag{3.37}$$

and from Eqs. (3.16) and (3.17)

$$\dot{\theta} = \frac{e}{r(n^2 r^2 - e^2)^{1/2}}$$

$$= \frac{e|\rho_0|}{r[n^2(1 - \rho_0 u)^2 - \bar{e}^2]^{1/2}}. \tag{3.38}$$

Thus, by noting Eqs. (3.9) and (3.33), we have

$$r\dot{\theta} = \frac{\varepsilon e|\rho_0|}{t} = \left(\frac{n_0}{t}\right)\sin\psi_0, \tag{3.39}$$

giving

$$r\dot{\alpha} = -\left(\frac{n_0}{t}\right)\cos(\psi_0 - \theta),$$

$$r\dot{\beta} = \left(\frac{n_0}{t}\right)\cos\theta. \tag{3.40}$$

Therefore, from Eqs. (3.34) and (3.35),

$$p = \frac{1}{1 - \rho_0 u}\left\{x_0|\rho_0|[\alpha t - n_0\cos(\theta - \psi_0)]\right.$$

$$\left. + \frac{p_0}{n_0}[\beta t + n_0\cos\theta]\right\}. \tag{3.41}$$

This and the corresponding equations for q and l can be written

$$p = A|\rho_0|x_0 + Bp_0,$$
$$q = A|\rho_0|y_0 + Bq_0, \tag{3.42}$$
$$l = Aa + Bl_0,$$

where

$$A = \frac{\alpha t - n_0\cos(\theta - \psi_0)}{1 - \rho_0 u},$$

$$B = \frac{\beta t/n_0 + \cos\theta}{1 - \rho_0 u}, \tag{3.43}$$

with t given by Eq. (3.33). However, with the help of Eq. (3.32), it is found that

$$\tan \psi = \frac{\overline{e}}{t}. \tag{3.44}$$

A useful check formula for numerical computations is provided by the relation

$$p^2 + q^2 + l^2 = n^2. \tag{3.45}$$

Another check formula is based on the well-known skewness invariant, i.e.,

$$xq - yp = x_0 q_0 - y_0 p_0, \tag{3.46}$$

which holds in any system having rotational symmetry about the z-axis. As a verification of the ray-tracing formulas derived above, it is useful to carry out the algebra to show that the tracing formulas do, indeed, satisfy Eqs. (3.45) and (3.46). This is done in Appendix B.

3.4 Technical Considerations

Before summarizing the ray-tracing procedure it should be noted that the formulas involve the assumption that the penetration depth u is measured from a sphere S_0 of constant index passing through the initial point P_0 of the ray. Since this sphere is not the same for different choices of the point P_0, it is advisable to establish, from the known index function, a formula for the penetration distance measured from some fixed reference sphere S_{00} concentric with S_0. This reference sphere might, for example, be chosen to pass through the vertex of the refracting surface being considered. It is then a simple matter to obtain the pertinent values of $n(u)$, with u measured from S_0, as required by the formulas.

Furthermore, it should be noted that the z-coordinate $\overline{z}$ is measured from the projection O' of P_0 on the optical axis (see Fig. 3.1). Consequently, since the position of O' is dependent on that of P_0, each $\overline{z}$ value must eventually be adjusted to give a distance from some fixed reference point in the optical system.

The given values for the ray trace include x_0, y_0, the index function $n(u)$, and the curvature ρ_0 of the equiindicial surface S_0 passing through P_0. Then $\varepsilon = \pm 1$ as $\rho_0 \gtrless 0$, corresponding to Figs. 3.1 and 3.2, respectively. Also required are the initial refractive index $n_0 = n(0)$ and the optical direction cosines p_0, q_0, and l_0 of the ray at P_0. As noted above, the z-coordinate of P_0 must also be known relative to a suitable reference point.

The constants a and b can be computed from Eqs. (3.30). The fact that b, as defined there, is the same as $n_0 \cos \psi_0$ is easily verified by observing that the direction cosines of OP_0 in Fig. 3.1 are x_0/r_0, y_0/r_0, and $a = \cos \alpha_0$, and those of the ray tangent are $(p_0, q_0, l_0)/n_0$, from which the last part of Eqs. (3.30) is justified. Knowing $\cos \psi_0$, we get $\bar{e}$ from Eq. (3.25).

Once values have been found for x_0, y_0, p_0, q_0, l_0, ρ_0, n_0, a, b, ε, $\bar{e}$, and ψ_0, the ray trace can be started. Given any value of $\bar{z}$, Eq. (3.29) is solved, in combination with Eqs. (3.24), (3.26), and (3.30), for u by the iterative scheme described above. By determining at the same time the corresponding values of α and M sinc θ, one can at once compute x and y from the first two of Eqs. (3.27).

With u already known, the corresponding value of n can be found, and accordingly the value of t given by Eq. (3.33). If, in the iteration process, values of β and θ have also been retained, it is easy to calculate A and B from Eqs. (3.43) and p, q, and l from Eqs. (3.42). Equations (3.45) and (3.46) provide useful check formulas.

It is understood that the formulas developed in the present section apply to most of the ray tracing met in practical situations. In unusual situations peculiarities can occur such as rays turning backward, so that the formulas would have to be reexamined carefully to deal with such cases. Throughout much of the book we consider only forward-moving rays ($l > 0$). In the next section we collect the tracing formulas for a spherical gradient.

3.5   Summary

a.   Given Quantities

(1)   Coordinates of Initial Point P_0. x_0, y_0, and z-coordinate of P_0 relative to a convenient fixed origin.

(2) Index Function. $n(u)$, index function expressed as a function of the penetration depth u relative to a sphere S_0 of constant index, $n_0 = n(0)$, passing through P_0 with curvature ρ_0 ($\rho_0 \gtrless 0$ as S_0 is concave or convex to the right for rays entering from the left).

(3) Initial Ray Direction. p_0, q_0, l_0, optical direction cosines with

$$p_0^2 + q_0^2 + l_0^2 = n_0^2. \tag{3.47}$$

(4) Terminal Point. $\bar{z}$, z-coordinate of P relative to O' (projection of P_0 on the z-axis).

b. Preliminary Calculations[†]

$$\begin{aligned}
\varepsilon &= \pm 1 \quad \text{as} \quad \rho_0 \gtrless 0, \qquad \varepsilon = +1 \quad \text{if} \quad \rho_0 = 0, \\
a &= \varepsilon[1 - \rho_0^2(x_0^2 + y_0^2)]^{1/2} = \cos \alpha_0, \\
b &= al_0 + |\rho_0|(p_0 x_0 + q_0 y_0) = n_0 \cos \psi_0, \\
e &= n_0 \sin \psi_0.
\end{aligned} \tag{3.48}$$

c. Computation of u

(1) Select a trial value u_1 of u. One choice is 0, but a better one is $\bar{z}/a$.

(2) Compute

$$M = \int_0^{u_1} \frac{du}{(1 - \rho_0 u)[n^2(1 - \rho_0 u)^2 - \bar{e}^2]^{1/2}} \tag{3.49}$$

and

$$\theta = \bar{e}|\rho_0|M. \tag{3.50}$$

(3) Compute a new trial value u_2 from

$$u_2 = \varepsilon \sec \theta \left\{ \frac{\bar{z}}{a} + M\left[(1 - \rho_0 u_1)\left(b - \frac{l_0}{a} \right) \right.\right.$$
$$\left.\left. + \bar{e} \sin\left(\frac{\theta}{2}\right) \operatorname{sinc}\left(\frac{\theta}{2}\right) \right] \right\}. \tag{3.51}$$

[†] See Eqs. (3.25) and (3.30).

(4)   Iterate, i.e., replace u_1 by u_2 and repeat. Continue iteration to determine limiting value of u. Retain, for later use, limiting values of M, $1 - \rho_0 u$, and other parameters.

d.   Coordinates of P

$$\beta = n_0 |\rho_0| M \operatorname{sinc} \theta,$$

$$\alpha = \cos \theta - \beta \frac{b}{n_0},$$

(3.52)

[from Eqs. (3.26) and (3.30)] and

$$x = (1 - \rho_0 u)(\alpha x_0 + p_0 M \operatorname{sinc} \theta),$$

$$y = (1 - \rho_0 u)(\alpha y_0 + q_0 M \operatorname{sinc} \theta).$$

(3.53)

e.   Direction of Ray at P

$$n = n(u),$$

$$t = \varepsilon \left[n^2 (1 - \rho_0 u)^2 - \bar{e}^2 \right]^{1/2},$$

$$A = \frac{\alpha t - n_0 \cos(\theta - \psi_0)}{1 - \rho_0 u},$$

$$B = \frac{\beta t / n_0 + \cos \theta}{1 - \rho_0 u},$$

(3.54)

and

$$p = A|\rho_0|x_0 + Bp_0,$$

$$q = A|\rho_0|y_0 + Bq_0,$$

$$l = Aa + Bl_0.$$

(3.55)

f.   Check Formulas

$$p^2 + q^2 + l^2 = n^2,$$

$$xq - yp = x_0 q_0 - y_0 p_0.$$

(3.56)

g.   Paraxial Rays

The tracing formulas for paraxial rays are obtained from the above exact formulas by neglecting quadratic and higher terms in x, y, p, q, and ψ_0 relative to unity.

This leads to the formulas

$$M = \int_0^z \frac{du}{n(1 - \rho_0 u)^2},$$

$$z = \bar{z},$$
$$x = (1 - \rho_0 z)(\alpha x_0 + M p_0),$$
$$y = (1 - \rho_0 z)(\alpha y_0 + M q_0), \qquad (3.57)$$

$$p = \frac{1}{1 - \rho_0 z}\left[(\alpha t - b)|\rho_0|x_0 + (1 + |\rho_0|Mt)p_0\right],$$

$$q = \frac{1}{1 - \rho_0 z}\left[(\alpha t - b)|\rho_0|y_0 + (1 + |\rho_0|Mt)q_0\right],$$

$$l = n,$$

where

$$\alpha = 1 + n_0 \rho_0 M,$$
$$t = \varepsilon n(1 - \rho_0 z), \qquad n = n(z), \qquad (3.58)$$
$$b = \varepsilon n_0, \qquad n_0 = n(0).$$

Here z is typically measured from the vertex of an object surface on which P_0 lies.

3.6   Numerical Example

It is helpful to have a simple example not only to illustrate the general formulas but also to check computer programs. For this purpose we examine again the Luneburg lens, for which the necessary quadrature can be performed in closed form. For simplicity we assume $r_0 = 1$ in

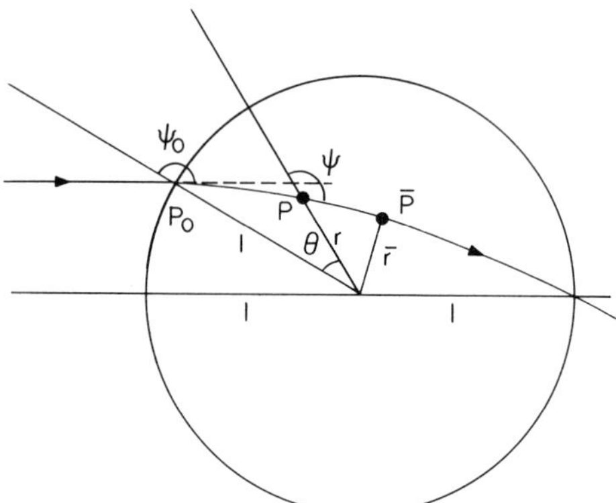

Fig. 3.4. Illustration of numerical example.

Eq. (2.24), in which case, with

$$n = (2 - r^2)^{1/2} \tag{3.59}$$

we have $n_0 = 1$ if we assume P_0 to lie on the surface of the lens (see Fig. 3.4).

In particular, we consider a ray entering horizontally from the left, with $\psi_0 = 150°$, as shown in the figure, and with P_0 located at the point of incidence on the lens surface. For the ray chosen

$$\varepsilon = -1, \qquad e = -\frac{1}{2}, \qquad \bar{e} = \sin \psi_0 = \frac{1}{2},$$

$$\cos \psi_0 = -\left(\frac{1}{2}\right)\sqrt{3}. \tag{3.60}$$

To find the angle ψ and the parameter t at a general point, we proceed as follows. From Eqs. (3.9) and (3.59)

$$\sin \psi = -\frac{e}{r(2 - r^2)^{1/2}} = \frac{1/2}{r^2(2/r^2 - 1)^{1/2}}. \tag{3.61}$$

Let

$$v = \frac{1}{r^2}, \tag{3.62}$$

giving

$$\sin \psi = \frac{v}{2(2v - 1)^{1/2}},\tag{3.63}$$

leading to

$$\cos \psi = -\frac{1}{2}\left[\frac{8v - v^2 - 4}{2v - 1}\right]^{1/2}\tag{3.64}$$

and

$$\tan \psi = -\frac{v}{(8v - v^2 - 4)^{1/2}}.\tag{3.65}$$

Then t is given by Eq. (3.44), i.e.,

$$t = \frac{1}{2\tan\psi}.\tag{3.66}$$

To find θ we introduce v as variable of integration in Eq. (3.8). Thus

$$\begin{aligned}
\theta &= e\int_1^r \frac{dr}{r(n^2 r^2 - e^2)^{1/2}} \\
&= -\int_1^r \frac{dr}{r^3[4(2/r^2 - 1) - 1/r^4]^{1/2}} \\
&= \frac{1}{2}\int_1^v \frac{dv}{(8v - v^2 - 4)^{1/2}}.
\end{aligned}\tag{3.67}$$

Integration then leads to

$$\theta = \frac{\pi}{6} + \tilde{\theta},\tag{3.68}$$

where

$$\tilde{\theta} = \frac{1}{2}\sin^{-1}\left[\frac{v - 4}{2\sqrt{3}}\right].\tag{3.69}$$

Introducing r from Eq. (3.62) we find

$$r = \left[4 + 2\sqrt{3}\sin\left(2\theta - \frac{\pi}{3}\right)\right]^{-1/2}, \tag{3.70}$$

as the polar equation of the ray.

There are several special points of interest on the ray. First, we note that $\tilde{\theta}$ vanishes and changes sign at $v = 4$, i.e., at $r = 1/2$. At that point $\theta = \pi/6$. A second point of interest is the stationary point $\bar{P}$, where $dr/d\theta = 0$ and $\dot{\theta} = \infty$. To determine the location of this point, we set

$$8v - v^2 - 4 = 0, \tag{3.71}$$

from which

$$v = 4 \pm 2\sqrt{3} = (\sqrt{3} \pm 1)^2. \tag{3.72}$$

Since $v = 1/r^2 \geq 1$, it is clear that the positive sign must be chosen. Hence,

$$\bar{v} = 4 + 2\sqrt{3} = (\sqrt{3} + 1)^2. \tag{3.73}$$

From this it follows that

$$\bar{r} = \frac{1}{2}(\sqrt{3} - 1). \tag{3.74}$$

Evidently at this point $\bar{\psi} = \pi/2$.

The value of θ at $\bar{P}$ is obtained from Eqs. (3.68), (3.69), and (3.73), the result being

$$\bar{\theta} = \frac{\pi}{6} + \frac{\pi}{4} = \frac{5\pi}{12}. \tag{3.75}$$

Thus $2\bar{\theta} = 150°$, showing that the ray, as expected, meets the opposite surface of the lens exactly on the axis.

3.7 Other Examples

As already mentioned, the Luneburg lens has limited usefulness because of the difficulty of making it and because the image surface is curved and located on the surface of the lens. However, by a slight generalization of Eq. (3.59), an index function is obtained having two parameters while still allowing exact evaluation of the integral in Eq. (3.8). Although the imaging obtained is no longer sharp, the lens designer is provided with two degrees of freedom whereby he can attempt to select a form that meets practical constraints and also gives a reasonable correction of aberrations.

We consider functions of the form

$$n = N_0(2 - \delta r^2)^{1/2}, \tag{3.76}$$

where again we take $r_0 = 1$ for simplicity. Actually, retaining r_0 as a parameter could be useful to give further flexibility in the design process. In this example Eq. (3.8) can be written

$$\theta = e \int_1^r \frac{dr}{r^3[(2 - \delta r^2)N_0^2/r^2 - e^2/r^4]^{1/2}}. \tag{3.77}$$

Setting

$$v = \frac{1}{r^2}, \qquad dv = -\frac{2}{r^3} dr \tag{3.78}$$

gives

$$\theta = \frac{\hat{e}}{2} \int_v^1 \frac{dv}{[2v - \delta - \hat{e}^2 v^2]^{1/2}} \tag{3.79}$$

where $\hat{e} = e/N_0$. This can be written

$$\theta = \frac{\hat{e}}{2A} \int_v^1 \frac{dv}{[1 - (\hat{e}/A)^2(v - 1/\hat{e}^2)^2]^{1/2}} \tag{3.80}$$

with

$$A = \frac{(1 - \delta \hat{e}^2)^{1/2}}{|\hat{e}|}. \tag{3.81}$$

Here it is clear that we must assume

$$1 - \delta \hat{e}^2 \geq 0. \tag{3.82}$$

The quadrature is now easily performed, the result being

$$\theta = \frac{1}{2}\,\varepsilon\left[\sin^{-1}\frac{1-\hat{e}^2/r^2}{(1-\delta\hat{e}^2)^{1/2}} - \sin^{-1}\frac{1-\hat{e}^2}{(1-\delta\hat{e}^2)^{1/2}}\right], \qquad (3.83)$$

where $\varepsilon = \pm 1$ as $dr/d\theta \gtrless 0$.

Chapter 4
AXIAL GRADIENTS

4.1 Introduction

If the refractive index is a function only of the distance z from a fixed plane, it is called an "axial" gradient. Usually the reference plane is perpendicular to the optical axis, making z a distance measured parallel to the axis, and hence the name for this type of gradient. Such a gradient can be formed by applying the ion-diffusion treatment to a plane glass surface. Once the gradient has been created, the surface can be ground into a curved shape.

With respect to lens design applications, Sands (1970) showed that the contributions of an axial gradient to the third-order aberrations of an optical system are equivalent to those of an aspheric surface.

Another application of axial gradients involves the use of an inhomogeneous layer formed at the surface of a plane sheet of glass, film, or metal to serve as an antireflection coating.

4.2 Equations of the Rays

An axial gradient can be regarded as the limiting case of a spherical gradient when the center of symmetry moves off to infinity. The corresponding ray equations can be obtained, therefore, from the formulas in the last chapter by taking $\rho_0 = 0$ and $\varepsilon = +1$. However, it is not difficult to derive the equations directly. Inspection of Eq. (1.7) with n a function of z alone shows that

$$p = nx' = p_0, \qquad q = ny' = q_0, \tag{4.1}$$

that is, the first two optical direction cosines are invariant along any ray. The third optical direction cosine is then obtained as a function of z by

$$l = (n^2 - p^2 - q^2)^{1/2} = (n^2 - p_0^2 - q_0^2)^{1/2}, \tag{4.2}$$

with n a given function of z. Here, as almost everywhere in this book, we assume for simplicity that $l > 0$.

To find the path of the ray we note that

$$p_0 = p = nx' = n\frac{\dot{x}}{\dot{s}} = l\dot{x} = l\frac{dx}{dz}, \tag{4.3}$$

and likewise for q_0. Therefore

$$x = x_0 + p_0 \int_{z_0}^{z} \frac{dz}{l},$$
$$\tag{4.4}$$
$$y = y_0 + q_0 \int_{z_0}^{z} \frac{dz}{l},$$

with l determined as a function of z by Eq. (4.2). For paraxial rays we note from Eq. (4.2) that we merely replace l by n in Eqs. (4.4). The optical path length along a ray in an axial gradient is given by

$$L = \int_{s_0}^{s} n\,ds = \int_{z_0}^{z} n\dot{s}\,dz = \int_{z_0}^{z} n^2 \frac{dz}{l}. \tag{4.5}$$

Since an axial gradient can be regarded as the limiting case of a spherical gradient when the center of symmetry moves away to infinity, it is expected that an invariant such as that given by Eq. (2.9) should exist for an axial gradient. This is, in fact, the case. From Eq. (4.2) we have

$$n\sin\psi = n(1 - \cos^2\psi)^{1/2}$$
$$= (n^2 - l^2)^{1/2} = (p^2 + q^2)^{1/2}$$
$$= (p_0^2 + q_0^2)^{1/2} = n_0\sin\psi_0, \tag{4.6}$$

where ψ is the angle the ray makes with the positive z-axis. This relation represents a generalization of Snell's refraction law.

4.3 A Special Axial Gradient

The determination of ray paths by means of Eqs. (4.4) should present no difficulties in cases where n is a given function of z, although, ordinarily, the quadrature involved must be carried out numerically. In certain cases, however, the integral can be evaluated in closed form. For example, if

$$n = N_0(1 + \alpha z)^{1/2}, \tag{4.7}$$

where the dielectric constant n^2 is a linear function of z, the exact integration is very simple. If we take note of Eq. (4.2) and introduce l as a new variable of integration, the result is

$$\begin{aligned}
x_1 &= x_0 + 2p_0 \left(\frac{l_1 - l_0}{N_0^2 \alpha} \right), \\
y_1 &= y_0 + 2q_0 \left(\frac{l_1 - l_0}{N_0^2 \alpha} \right).
\end{aligned} \tag{4.8}$$

Here subscripts 0 and 1 indicate values at any two points P_0 and P_1 on the ray. An alternative form of these equations is obtained by multiplying and dividing by $l_1 + l_0$ in each of the last terms and evaluating l_1^2 and l_0^2 with the help of Eqs. (4.2) and (4.7). The result is

$$\begin{aligned}
x_1 &= x_0 + 2p_0 \left(\frac{z_1 - z_0}{l_1 + l_0} \right), \\
y_1 &= y_0 + 2q_0 \left(\frac{z_1 - z_0}{l_1 + l_0} \right).
\end{aligned} \tag{4.9}$$

4.4 Wave Optics Considerations

It has been known for a long time that a suitable inhomogeneous layer placed on a plane surface substantially reduces the reflectivity. Actually the same principle can be applied to a curved surface, but for simplicity and in the context of the present chapter we consider only the case of a layer consisting of an axial gradient.

The analysis of this problem requires the use of wave optics. Abelès (1950), Jacobsson (1963), Born and Wolf (1970), and others have described the theory, the subject usually being referred to as the propagation of light through stratified media.

In the case of a homogeneous layer with plane parallel faces, it is known that the effect of a plane polarized plane wave incident on the layer is determined by a matrix M of four elements indicated by

$$M = \begin{bmatrix} m_{11} & m_{12} \\ m_{21} & m_{22} \end{bmatrix}. \tag{4.10}$$

The elements m_{ij} depend on the optical constants ε and μ of the medium, the thickness d of the layer, the inclination angle ψ of the plane wave within the layer, the direction of polarization of the incident wave, and the wavelength λ of the light.

When the quantities m_{ij} are known, the reflectivity coefficient r is given by the formula

$$r = \frac{(m_{11} + m_{12}u_s)u_a - (m_{21} + m_{22}u_s)}{(m_{11} + m_{12}u_s)u_a + (m_{21} + m_{22}u_s)}. \tag{4.11}$$

Here the values u_a and u_s depend on the direction of propagation and the direction of polarization of the wave as described below. The reflectivity R of the layer is then given by

$$R = |r^2|. \tag{4.12}$$

For natural (unpolarized) light the reflectivity is obtained as follows. Equations (4.10), (4.11), and (4.12) are evaluated for a TE-wave (electric vector perpendicular to the plane of incidence) and a TM-wave (electric vector parallel to the plane of incidence), giving reflectivities R_1 and R_2 respectively. Then the simple average

$$R = \frac{1}{2}(R_1 + R_2) \tag{4.13}$$

gives the reflectivity for natural light.

The quantities u_a and u_s in Eq. (4.11) are defined as follows. We define u by

$$u = \left(\frac{\varepsilon}{\mu}\right)^{1/2} \cos\psi \quad \text{or} \quad u = \left(\frac{\mu}{\varepsilon}\right)^{1/2} \cos\psi, \tag{4.14}$$

depending on whether the wave is the TE-wave or the TM-wave. The subscripts a and s in Eq. (4.11) refer to the media just before and just after the layer, respectively. Owing to Eq. (4.6) and the ordinary refraction law, the quantity

$$(\varepsilon\mu)^{1/2} \sin \psi = n \sin \psi = s \tag{4.15}$$

is the same constant, for a given plane wave, in all three media. Thus, if the inclination angle ψ_a of the incident wave is given, the value of ψ within the layer and its value ψ_s following the layer are easily determined.

We now consider a stratified medium with plane parallel faces but with ε and μ varying continuously with the distance z from the first surface (axial gradient). A classical method of analyzing such a layer is to approximate it by a stack of N homogeneous layers, each of thickness Δz, so that a matrix M_k can be deduced for each. Then an effective matrix for the entire stack is found by simply multiplying the individual matrices. If N is large and Δz small, the resulting matrix M serves as a good approximate matrix for the original stratified medium. In fact, by passing to the limit as $N \to \infty$ and $\Delta z \to 0$, we can deduce exact formulas for the elements of the effective matrix M for the stratified layer.

The above approach has proved useful in many applications. Also it serves as a basis for proving several general laws for stratified layers. For example, it is easily shown that Eq. (4.15) is still valid for a given plane wave passing through the layer even though the angle ψ as well as ε, μ, and n vary continuously in the interior of the layer (cf. Eq. (4.6)).

The approach described above leads to somewhat complicated formulas involving integrals that can be evaluated exactly only in special cases. We therefore consider an alternative way of analyzing stratified layers, i.e., the Wentzel–Kramers–Brillouin–Jeffries or WKBJ method. Actually, this method was first given by Liouville in 1837 and later rediscovered by Lord Rayleigh in 1912 and by Gans in 1915.

In the WKBJ approach an attempt is made to solve Maxwell's differential equations directly for the case of a stratified layer. The essential idea of the method is to introduce, in place of z, a new independent variable

$$\theta(z) = k \int_0^z (\varepsilon\mu)^{1/2} \cos \psi \, dz, \tag{4.16}$$

where $k = 2\pi/\lambda$. The geometrical meaning of θ is simply the optical path length within the layer, converted to units of phase.

When the above substitution is made in the differential equations, it is seen that two small terms can be neglected provided the wavelength is small. The resulting equations are then easily solved. Thus the WKBJ method produces simple, approximate formulas that prove to be very reliable when λ is small, as in the case of visible light.

With λ small, k is large, and so inspection of Eq. (4.16) suggests that the WKBJ method tends to work best when $n = (\varepsilon\mu)^{1/2}$ is a relatively slowly changing function of z.

As shown by Jacobsson (1963), the application of this method leads to the following formula for the reflection coefficient for a stratified layer of thickness d:

$$r = \frac{c_0[\beta_s\beta(0) - \beta_a\beta(d)] + is_0[\beta(0)\beta(d) - \beta_a\beta_s]}{c_0[\beta_s\beta(0) + \beta_a\beta(d)] + is_0[\beta(0)\beta(d) + \beta_a\beta_s]}. \tag{4.17}$$

Here

$$c_0 = \cos\theta(d), \qquad s_0 = \sin\theta(d), \tag{4.18}$$

and

$$\frac{1}{\beta(z)} = \left[\frac{\varepsilon(z)}{\mu(z)}\right]^{1/2} \cos\psi(z) \tag{4.19}$$

for the TE-wave, and

$$\frac{1}{\beta(z)} = \left[\frac{\mu(z)}{\varepsilon(z)}\right]^{1/2} \cos\psi(z) \tag{4.20}$$

for the TM-wave. As before, the subscripts a and s in Eq. (4.17) indicate values just before and just after the layer.

4.5 Axial Gradient as an Antireflection Coating

We consider the problem of reducing the reflectivity at a plane surface in air. The base material could be glass, plastic, emulsion, or other substance, assumed for now to be a dielectric (transparent and nonmagnetic). In this case, in the base material,

$$n_s = \sqrt{\varepsilon_s}, \qquad \mu_s = 1 \tag{4.21}$$

with ε_s and n_s real. Suppose now that, in the region near the air surface (at $z = 0$), the medium is modified so that its refractive index rises slowly from the value n_0 at $z = 0$ to the base value n_s at a distance $z = d$.

With a transition layer of the above type there is no discontinuity of the refractive index at the second surface, and therefore no reflection can occur there. This fact is beneficial to the objective of reducing the overall reflectivity and also leads to simplification in the reflectivity formula.

In the present case $\beta(d) = \beta_s$, so that Eq. (4.17) reduces to

$$r = \frac{\beta(0) - \beta_a}{\beta(0) + \beta_a}. \tag{4.22}$$

Here, from Eqs. (4.19), (4.20), and (4.15),

$$\beta(0) = (n_0^2 - s^2)^{-1/2} \tag{4.23}$$

for a TE-wave and

$$\beta(0) = n_0^2(n_0^2 - s^2)^{-1/2} \tag{4.24}$$

for a TM-wave, with

$$\beta_a = (1 - s^2)^{-1/2}, \qquad s = \sin\psi_a \tag{4.25}$$

holding in both cases. Equation (4.12) gives, in either case,

$$R = \left(\frac{a-1}{a+1}\right)^2, \qquad a = \frac{\beta(0)}{\beta_a} \tag{4.26}$$

for the reflectivity. For natural (unpolarized) light Eq. (4.26) should be evaluated for the TM- and TE-waves separately and the results averaged.

Inspection of the preceding formulas reveals that the reflectivity in the presence of the transition layer is essentially independent of the thickness of the layer and also of the profile of the index function. However, the approximation inherent in the WKBJ formulation requires that the layer be many wavelengths thick and also that the index function not change rapidly within the layer. The reflectivity depends on the index value n_0 of the layer at the outer surface but not on the index of the base material. It will be seen presently that the smaller the value of n_0, the lower the reflectivity tends to be.

In order to estimate the effectiveness of this type of transition layer or coating, a comparison can be made with the reflectivity obtained without the layer, i.e., at a simple plane surface with air on one side and material with base index n_s on the other. Here the reflectivity is given by the well-known Fresnel formulas

$$R_1 = \frac{\sin^2(\psi_a - \psi_s)}{\sin^2(\psi_a + \psi_s)} \tag{4.27}$$

and

$$R_2 = \frac{\tan^2(\psi_a - \psi_s)}{\tan^2(\psi_a + \psi_s)} \tag{4.28}$$

for the TE-wave and the TM-wave respectively.

With the incident beam in air

$$n_s \sin \psi_s = \sin \psi_a = s. \tag{4.29}$$

At normal incidence ($s = 0$), for the case with the transition layer present, Eq. (4.26) gives

$$R = \left(\frac{n_0 - 1}{n_0 + 1}\right)^2 \tag{4.30}$$

for both waves, and so also for natural light. With $s = 0$, Eqs. (4.27) and (4.28) both reduce to

$$R = \left(\frac{n_s - 1}{n_s + 1}\right)^2. \tag{4.31}$$

As an example, with $n_0 = 1.4$, the value of R given by Eq. (4.30) is 0.028, for the reflectivity in the presence of the transition layer. If $n_s = 1.6$, Eq. (4.31) gives $R = 0.053$, showing that the reflectivity at normal incidence is nearly twice as great when the transition layer is not present. Similar comparisons can be made for nonnormal incidence ($s \neq 0$) based on Eqs. (4.26), (4.27), and (4.28).

It is, of course, well known that reflectivity can be greatly reduced by means of a quarter-wave homogeneous layer as a coating on a plane surface. This procedure, however, has certain drawbacks. The layer must be very thin, the optical thickness nd being only one quarter of the wavelength in the simplest case. Also, control of the thickness is critical since the action of the layer depends on interference. Furthermore, the reflectivity is wavelength dependent, whereas this is not appreciably so in the case of a gradient transition layer.

As with interference coatings, gradient layers can be used to increase reflectivity instead of decreasing it. In this case the index n_0 of the layer at its outer surface should be large instead of small.

If the base material is metallic, its refractive index is a complex number, so that a true transition layer cannot be made of a transparent material. However, a dielectric gradient layer can still serve to reduce the reflectivity substantially. We study this possibility for the case of normal incidence, starting again with Eq. (4.17) and from Eq. (4.20),

$$\beta(0) = n_0, \qquad \beta(d) = n_1,$$
$$\beta_a = 1, \qquad \beta_s = \alpha + i\beta. \tag{4.32}$$

Here β_s is the complex refractive index of the base material, n_0 and n_1 are the initial and terminal index values of the gradient layer, and d is its thickness. Here

$$c_0 = \cos \theta(d), \qquad s_0 = \sin \theta(d),$$
$$\theta(d) = k \int_0^d n(z)\, dz, \qquad k = \frac{2\pi}{\lambda}, \tag{4.33}$$

$n(z)$ being the index function of the layer.

Substitution of these values into Eq. (4.17) gives

$$r = \frac{\left[(\alpha + i\beta)n_0 - n_1\right] + it_0\left[n_0 n_1 - (\alpha + i\beta)\right]}{\left[(\alpha + i\beta)n_0 + n_1\right] + it_0\left[n_0 n_1 + (\alpha + i\beta)\right]}, \tag{4.34}$$

where $t_0 = s_0/c_0$. Separating the real and imaginary parts in the numerator and denominator, we find the reflectivity given by

$$R = |r|^2 = \frac{a^2 + b^2}{c^2 + d^2} \tag{4.35}$$

with

$$a = \alpha + \frac{\beta t_0 - n_1}{n_0}, \qquad b = \beta + t_0 \frac{(n_1 - \alpha)}{n_0},$$
$$c = \alpha - \frac{\beta t_0 - n_1}{n_0}, \qquad \overline{d} = \beta + t_0 \frac{(n_1 + \alpha)}{n_0}. \tag{4.36}$$

It is known that the exact profile of the index function is not critical in problems of this kind. If a linear function

$$n(z) = n_0 + (n_1 - n_0)\frac{z}{d} \qquad (4.37)$$

is assumed, it follows that

$$\theta(d) = \frac{kd(n_0 + n_1)}{2}. \qquad (4.38)$$

In choosing the gradient parameters, it is advantageous, as might be expected, to have the external index value n_0 as small as possible. It is usually not feasible to match the inner index value n_1 to the absolute value $(\alpha^2 + \beta^2)^{1/2}$ of the complex base index. However, the next best thing is to choose n_1 as large as possible so as to minimize the reflection at the interface.

In the case of a metallic surface with no coating, the reflectivity is found from Eq. (4.31), which must now be written with absolute value signs and with n_s given as a complex number. Thus

$$R = \left|\frac{n_s - 1}{n_s + 1}\right|^2 = \left|\frac{\alpha + i\beta - 1}{\alpha + i\beta + 1}\right|^2$$
$$= \frac{(\alpha - 1)^2 + \beta^2}{(\alpha + 1)^2 + \beta^2}. \qquad (4.39)$$

With selenium, for example,

$$n_s = 4.4 + 0.36i, \qquad (4.40)$$

resulting in a value of $R = 0.4$. This means that, at normal incidence, the reflectivity is 40%, and would be substantially higher for nonnormal incidence. If a transparent linear gradient coating is applied with $n_0 = 1.38$, $n_1 = 2.24$, $d = 3\mu$, and $\lambda = 0.6\mu$, Eqs. (4.35) and (4.36) give a value of $R = 0.206$. The coating reduces the reflectivity by almost half.

Perhaps the values assumed above cannot be achieved by available materials, but it is certainly feasible to reduce reflectivity considerably by means of gradient coatings. Jacobsson (1968) describes several methods for producing such coatings. These mostly involve modifications of the vapor-deposition technique.

Further discussion and details on the use of a gradient coating to reduce reflectivity are given by Ritchie and Window (1977), where the chief interest is in the application to solar-energy absorbers.

Chapter 5
RADIAL GRADIENTS

5.1 Basic Equations

The term "radial" gradient is used for an index distribution having cylindrical symmetry, n being a function of the radial distance r from a fixed line. In this case it is appropriate to employ cylindrical coordinates (r, θ, z) with the z-axis as the axis of symmetry, i.e.,

$$x = r\cos\theta, \qquad y = r\sin\theta, \qquad z = z. \tag{5.1}$$

The reader should note that r and θ are now not defined the same as in earlier chapters.

For the analysis of radial gradients it is convenient first to use Eq. (1.7). Since $\partial n/\partial z = 0$, the third component of Eq. (1.7) shows at once that

$$nz' = n\frac{dz}{ds} = l_0 = \text{const.} \tag{5.2}$$

Thus, with the optical direction cosines defined by

$$\begin{aligned}
p &= nx' = n\cos\alpha, \\
q &= ny' = n\cos\beta, \\
l &= nz' = n\cos\gamma,
\end{aligned} \tag{5.3}$$

Eq. (5.2) shows that the third optical direction cosine is invariant along any ray, $l = l_0$.

In cylindrical coordinates Eq. (1.2) takes the form

$$L = \int_{z_0}^{z} F\,dz \tag{5.4}$$

with

$$F = n(r)(1 + \dot{r}^2 + r^2\dot{\theta}^2)^{1/2}, \tag{5.5}$$

where a dot indicates d/dz and $dz/ds > 0$.

The Euler equation for θ then gives

$$\frac{d}{dz}\left[nr^2\dot{\theta}(1 + \dot{r}^2 + r^2\dot{\theta}^2)^{-1/2}\right] = 0, \tag{5.6}$$

showing that

$$nr^2\dot{\theta}(1 + \dot{r}^2 + r^2\dot{\theta}^2)^{-1/2} = c \tag{5.7}$$

is constant along any ray. Clearly c must have the same sign as $\dot{\theta}$. It will be seen later that c represents the well-known skewness invariant valid for any ray in any system having cylindrical symmetry.

Equation (5.7) can be solved for $\dot{\theta}$, the result being

$$\dot{\theta} = \frac{c(1 + \dot{r}^2)^{1/2}}{mr^2}, \tag{5.8}$$

where, following Luneburg (1964), we have introduced the function

$$m = \left(n^2 - \frac{c^2}{r^2}\right)^{1/2}. \tag{5.9}$$

The Euler equation for r is

$$\left(\frac{d}{dz}\right)\left[n\dot{r}(1 + \dot{r}^2 + r^2\dot{\theta}^2)^{-1/2}\right]$$

$$= (1 + \dot{r}^2 + r^2\dot{\theta}^2)^{1/2}\frac{dn}{dr}$$

$$+ nr\dot{\theta}^2(1 + \dot{r}^2 + r^2\dot{\theta}^2)^{-1/2}. \tag{5.10}$$

Using Eqs. (5.8) and (5.9) to eliminate $\dot{\theta}$ we find

$$(d/dz)\big[m\dot{r}(1 + \dot{r}^2)^{-1/2}\big] = (1 + \dot{r}^2)^{1/2}\frac{dm}{dr}. \tag{5.11}$$

Equation (5.11) can be solved in terms of a single quadrature. First, by writing it in the form

$$\big[m\dot{r}(1 + \dot{r}^2)^{-1/2}\big]\,d\big[m\dot{r}(1 + \dot{r}^2)^{-1/2}\big] = m\,dm, \tag{5.12}$$

we have

$$\big[m\dot{r}(1 + \dot{r}^2)^{-1/2}\big]^2 + k = m^2, \tag{5.13}$$

or

$$\frac{m^2}{1 + \dot{r}^2} = k, \tag{5.14}$$

where k is a constant of integration.

The optical significance of the above result is seen as follows. The third optical direction cosine l can be written

$$l = nz' = n/\dot{s} = n(1 + \dot{r}^2 + r^2\dot{\theta}^2)^{-1/2}, \tag{5.15}$$

if z is an increasing function of s.

From Eqs. (5.8) and (5.9) it follows that

$$l^2 = \frac{m^2}{1 + \dot{r}^2} = k = l_0^2. \tag{5.16}$$

This verifies again that l is invariant along a ray.

From Eq. (5.16) we have

$$\dot{r}^2 = \frac{m^2}{l_0^2} - 1 \tag{5.17}$$

or

$$\dot{r} = \frac{dr}{dz} = \frac{g}{l_0} \tag{5.18}$$

with g defined by

$$g = \pm(m^2 - l_0^2)^{1/2}. \tag{5.19}$$

Here the $\pm$ is chosen as r is an increasing or decreasing function of z. Integration now gives

$$z - z_0 = l_0 \int_{r_0}^{r} \frac{dr}{g}. \tag{5.20}$$

In principle, the paths of the rays are now completely determined. For Eq. (5.20) gives z as a function of r, provided n, and thereby m, is a known function of r. The resulting formulas can be solved, in principle, for r in terms of z.

To obtain θ as a function of z we use Eq. (5.16) to write Eq. (5.8) as

$$\dot{\theta} = \frac{d\theta}{dz} = \frac{c}{l_0 r^2}, \tag{5.21}$$

so that

$$\theta = \theta_0 + \frac{c}{l_0} \int_{z_0}^{z} \frac{dz}{r^2}. \tag{5.22}$$

With r previously determined as a function of z, Eq. (5.22) yields θ as a function of z. Once both r and θ are known as functions of z, Eqs. (5.1) give x and y in terms of z.

The optical direction cosines of the ray at any point can be obtained easily once the above equations have been evaluated. For

$$p = nx' = n\frac{\dot{x}}{\dot{s}} = l\dot{x} = l_0\dot{x},$$

$$q = ny' = n\frac{\dot{y}}{\dot{s}} = l\dot{y} = l_0\dot{y}, \tag{5.23}$$

$$l = l_0.$$

But, from Eqs. (5.1)

$$\dot{x} = \dot{r}\cos\theta - r\sin\theta\dot{\theta},$$
$$\dot{y} = \dot{r}\sin\theta + r\cos\theta\dot{\theta}. \tag{5.24}$$

Hence, by virtue of Eqs. (5.18) and (5.21),

$$p = g \cos \theta - \frac{c}{r} \sin \theta,$$

$$q = g \sin \theta + \frac{c}{r} \cos \theta, \tag{5.25}$$

$$l = l_0.$$

It can now be readily verified, by means of Eqs. (5.1) and (5.25), that c represents the skewness invariant, i.e.,

$$xq - yp = c = x_0 q_0 - y_0 p_0. \tag{5.26}$$

Equation (5.26) can be used to determine the value of c at the start of the trace of any ray. This relation also provides a useful check formula for numerical computations. Another useful check formula is

$$p^2 + q^2 + l^2 = n^2, \tag{5.27}$$

since n can be found as a function of z once r has been found in terms of z.

With r known in terms of z the light path L can be obtained by first determining n as a function of z and using Eq. (1.2) in the form

$$L = \int_{s_0}^{s} n\dot{s}\, dz = \int_{z_0}^{z} n^2 \frac{dz}{l} = (1/l_0) \int_{z_0}^{z} n^2\, dz. \tag{5.28}$$

5.2   Practical Considerations

In the last section it was shown that the paths and directions of rays in a radial gradient are completely determined by two quadratures. However, the equation giving r in terms of z is an implicit one, which usually can be solved only by numerical methods. Furthermore, the integrals involved become singular for certain points on certain rays. In view of these technical difficulties a number of authors have developed ray-tracing methods for radial gradients. Some of these methods will be described in the next chapter. However, we mention here one of the most basic approaches based on standard computer subroutines (see Marchand, 1972).

Differentiation of Eq. (5.17) leads to

$$\ddot{r} = \frac{m \, dm/dr}{l_0^2}, \tag{5.29}$$

a second-order differential equation between r and z with m defined by Eq. (5.9). If we let $v = \dot{r}$ and take note of Eqs. (5.21) and (5.9), we have

$$\dot{r} = v,$$

$$\dot{v} = \frac{m \, \partial m/\partial r}{l_0^2}, \tag{5.30}$$

$$\dot{\theta} = \frac{c}{l_0 r^2}.$$

This set of differential equations can be solved by standard computer subroutines for r, v, and θ as well as $\dot{r}$, $\dot{v}$, and $\dot{\theta}$ in terms of z, provided suitable initial conditions are given. For a meridional ray, $c = 0$, in which case the third equation can be ignored and $m = n$.

A simpler system is obtained merely by using Eqs. (5.17) and (5.21), i.e.,

$$\dot{r} = \pm \left(\frac{m^2}{l_0^2} - 1 \right)^{1/2},$$

$$\dot{\theta} = \frac{c}{l_0 r^2}. \tag{5.31}$$

But here it is somewhat awkward to apply a sign change whenever $\dot{r}$ changes sign.

A single differential equation between r and θ can be obtained by dividing the first of Eqs. (5.31) by the second, giving

$$\frac{dr}{d\theta} = \pm (m^2 - l_0^2)^{1/2} \frac{r^2}{c}, \tag{5.32}$$

the $\pm$ being chosen as $dr/d\theta \gtrless 0$. With the help of Eq. (5.9) we then have

$$\theta = \theta_0 \pm c \int_{r_0}^{r} \frac{dr}{r[r^2(n^2 - l_0^2) - c^2]^{1/2}}. \tag{5.33}$$

Here again the integral can become singular at certain points, and the sign changes require attention. Furthermore, the dependence of n and θ on z must still be determined.

Of these several approaches, Eqs. (5.30) appears to be the best for most practical purposes. Of course, once r and θ are found in terms of z, x and y are obtained by Eqs. (5.1) and the direction cosines by Eqs. (5.25).

5.3 A Special Radial Gradient

In view of the lack of a simple ray-tracing routine for radial gradient media, it is helpful to note special index functions for which the integrations can be carried out in closed form. An important example is that for which

$$n^2 = N_0^2 \pm b^2 r^2, \qquad b > 0. \tag{5.34}$$

In this case the dielectric constant n^2 is a quadratic function of r with no linear term present. It will be seen that for either choice of sign in Eq. (5.34) the differential equations for the rays can be integrated to give simple, exact ray-tracing formulas.

For the present example it is best to adopt Cartesian coordinates and to convert Eq. (1.7) to derivatives with respect to z instead of s. In a radial gradient we have

$$\frac{\partial n}{\partial z} = 0, \tag{5.35}$$

so that, in any radial gradient, Eqs. (1.8) can be reduced to the form

$$\ddot{\mathbf{r}} = \frac{n\,\nabla n}{l_0^2}. \tag{5.36}$$

The equation in this form is often inconvenient, for its component equations are two second-order differential equations that are coupled, both right-hand sides being functions of x, y, and r. However, assuming Eq. (5.34) with the negative sign, we have

$$n\,\nabla n = (n\,dn/dr)\frac{\mathbf{r}}{r} = -b^2\mathbf{r}. \tag{5.37}$$

In this case Eq. (5.36) reduces to

$$\ddot{\mathbf{r}} + \frac{\mathbf{r}b^2}{l_0^2} = 0. \tag{5.38}$$

If we modify the z-coordinate by putting

$$\bar{z} = \left(z - z_0\right)\frac{b}{l_0}, \tag{5.39}$$

we have

$$\frac{d^2x}{d\bar{z}^2} + x = 0,$$

$$\frac{d^2y}{d\bar{z}^2} + y = 0 \tag{5.40}$$

if the negative sign is present in Eq. (5.34). The solution of these equations is given by

$$x = x_0 \cos\bar{z} + \frac{p_0}{b}\sin\bar{z},$$

$$y = y_0 \cos\bar{z} + \frac{q_0}{b}\sin\bar{z}, \tag{5.41}$$

where x_0, y_0, p_0, q_0 are the values of x, y, p, and q respectively when $\bar{z} = 0$.

The optical direction cosines at a general point of the ray are found by differentiation of Eqs. (5.41), for

$$p = nx' = n\frac{\dot{x}}{\dot{s}} = l\dot{x} = l_0\dot{x} = b\frac{dx}{d\bar{z}}. \tag{5.42}$$

Hence

$$p = p_0 \cos\bar{z} - bx_0 \sin\bar{z},$$
$$q = q_0 \cos\bar{z} - by_0 \sin\bar{z}, \tag{5.43}$$
$$l = l_0.$$

When the positive sign is present in Eq. (5.34), the differential equations reduce to

$$\frac{d^2 x}{d\bar{z}^2} - x = 0,$$
$$\frac{d^2 y}{d\bar{z}^2} - y = 0,$$

$$(5.44)$$

leading to the solution

$$x = x_0 \cosh \bar{z} + \frac{p_0}{b} \sinh \bar{z},$$

$$y = y_0 \cosh \bar{z} + \frac{q_0}{b} \sinh \bar{z},$$

$$p = p_0 \cosh \bar{z} + b x_0 \sinh \bar{z},$$

$$q = q_0 \cosh \bar{z} + b y_0 \sinh \bar{z}.$$

$$(5.45)$$

It is not difficult to verify that the solutions in both cases satisfy the check formulas given by Eqs. (5.26) and (5.27).

In Eqs. (5.41) and (5.45) division by b can cause indeterminate results if b is small. However, this difficulty can be overcome by noting that (assuming for now that $z_0 = 0$ in Eq. (5.39))

$$\frac{1}{b} \sin \bar{z} = \frac{1}{b} \sin \frac{bz}{l_0}$$
$$= \frac{z}{l_0} \left(\frac{1}{z} \sin \bar{z} \right).$$

$$(5.46)$$

In numerical computation, when $\bar{z}$ is small (say $\bar{z} < 0.001$), the following series can be used:

$$\frac{1}{z} \sin \bar{z} = 1 - \bar{z}^2/3! + \bar{z}^4/5! - \cdots .$$

$$(5.47)$$

A similar device can be used in computing with Eq. (5.45) when $\bar{z}$ is small.

5.4   The GRIN Rod Problem

In designing gradient fiber elements one goal is to select an index profile such that images can be transmitted along a fiber. A cylindrical rod having a radial gradient has been called a GRIN (graded-index) rod. It appears that no ideal index function for exact imaging exists.

In a GRIN rod all meridional rays are sharply imaged if the index function has the form

$$n = N_0 \operatorname{sech}(\alpha r), \tag{5.48}$$

where N_0 and α are constants (Fletcher *et al.*, 1954). To verify this we note that $c = 0$ for meridional rays, so that Eq. (5.17) reduces to

$$\dot{r} = \pm \left(\frac{n^2}{l_0^2} - 1 \right)^{1/2} = \pm \left(\frac{N_0^2}{l_0^2} \operatorname{sech}^2(\alpha r) - 1 \right)^{1/2}, \tag{5.49}$$

with $\pm$ chosen as $\dot{r} \gtrless 0$.

This leads to

$$z = z_0 \pm \frac{1}{\alpha} \int_{\bar{r}_0}^{\bar{r}} \frac{\cosh \bar{r} \, d\bar{r}}{(A^2 - \sinh^2 \bar{r})^{1/2}}, \tag{5.50}$$

with

$$\bar{r} = \alpha r, \qquad A = \left(\frac{N_0^2}{l_0^2} - 1 \right)^{1/2}. \tag{5.51}$$

Defining a new variable of integration by

$$U = \sinh \bar{r} \tag{5.52}$$

leads to

$$\sinh(\alpha r) = A \sin \left[\sin^{-1} \frac{U_0}{A} \pm \alpha(z - z_0) \right], \qquad U_0 = \sinh(\alpha r_0). \tag{5.53}$$

Accordingly it is clear that every meridional ray repeats itself periodically (if the fiber is sufficiently long) and the period is independent of the ray or the starting point chosen. Hence, every fan of meridional rays starting at a given object point is sharply imaged periodically within the fiber.

Unfortunately, the index gradient specified by Eq. (5.48) does not image skew rays sharply. In connection with fiber optics studies the special case of helical rays has received some attention. These are skew rays each of which lies at a constant distance from the axis of symmetry, as shown in Fig. 5.1.

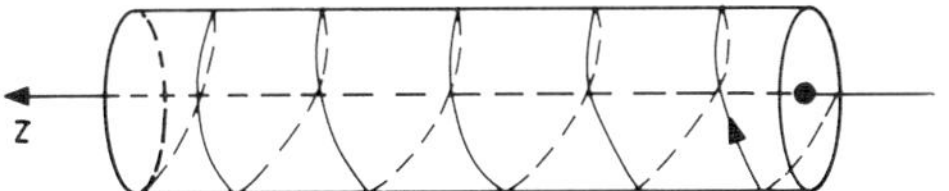

Fig. 5.1.   Helical ray in a radial gradient.

For a helical ray, we have

$$\ddot{r} = 0, \qquad \dot{r} = 0, \qquad r = r_0, \qquad n = n_0, \qquad \frac{dn}{dr} = \left(\frac{dn}{dr}\right)_0 \qquad (5.54)$$

at every point of the ray. From Eq. (5.7)

$$n_0 r_0^2 \dot{\theta}(1 + r_0^2 \dot{\theta}^2)^{-1/2} = c, \qquad (5.55)$$

showing that $\dot{\theta}$ is constant along the ray. Therefore

$$\theta = \theta_0 + \dot{\theta}_0(z - z_0) \qquad (5.56)$$

and

$$\begin{aligned}
x &= r_0 \cos[\theta_0 + \dot{\theta}_0(z - z_0)], \\
y &= r_0 \sin[\theta_0 + \dot{\theta}_0(z - z_0)].
\end{aligned} \qquad (5.57)$$

Thus a helical ray follows a periodic path, the period being given by $2\pi/\dot{\theta}_0$.

We shall see that, when the index function $n(r)$ and the initial radial distance r_0 have been specified, only two helical rays are possible. These are symmetrically placed with respect to the meridional plane through P_0. Ordinarily the value of $\dot{\theta}_0$ depends on both the function $n(r)$ and the value of n_0, so that helical rays starting at different heights usually have different periods. However, for certain types of index gradients, the value of $\dot{\theta}_0$ is independent of r_0, in which case all the helical rays have the same period.

From Eq. (5.29) with $\ddot{r} = 0$ it is seen that on a helical ray we must have

$$\frac{dm^2}{dr} = 0 \tag{5.58}$$

at all points. Hence, from Eq. (5.9),

$$n\frac{dn}{dr} + \frac{c^2}{r^3} = 0, \tag{5.59}$$

or

$$c^2 = -r^3 n\frac{dn}{dr}. \tag{5.60}$$

It is clear that, for c to be real, we must have $dn/dr < 0$. Thus a helical ray can exist only at a point where n is a decreasing function of r. Also, Eq. (5.60) verifies that just two symmetrical directions are possible for helical rays leaving a given point.

From Eqs. (5.31) with $\dot{r} = 0$, we have

$$l_0^2 = m^2 = n^2 - \frac{c^2}{r^2} \tag{5.61}$$

and

$$\dot{\theta}^2 = \frac{c^2}{l_0^2 r^4}, \tag{5.62}$$

giving

$$\dot{\theta}^2 = \frac{1}{r^2(n^2 r^2/c^2 - 1)}. \tag{5.63}$$

Now consider a gradient profile given by

$$n^2 = \frac{N_0^2}{1 + \beta^2 r^2}. \tag{5.64}$$

Then

$$n\,dn/dr = \frac{-N_0^2\beta^2 r}{(1 + \beta^2 r^2)^2}$$

$$= -\frac{n^4\beta^2 r}{N_0^2}, \tag{5.65}$$

from which

$$c^2 = \frac{r^4 n^4 \beta^2}{N_0^2}. \tag{5.66}$$

Substitution into Eq. (5.63) than gives

$$\dot{\theta}^2 = \beta^2 \tag{5.67}$$

and

$$\dot{\theta}_0 = \pm\beta. \tag{5.68}$$

For index functions having the form of Eq. (5.64), all helical rays have the same period. This, however, does not mean that any point is sharply imaged, for only two helical rays can start from any given point. There are, in fact, many nonhelical skew rays starting at any given object point.

The special function discussed in the last section likewise does not provide an ideal solution for the GRIN rod problem. If the negative sign is chosen in Eq. (5.34), it is seen that all rays are periodic whether paraxial, skew, or meridional. As seen from Eqs. (5.39), (5.41), and (5.43), the period of each ray is given by

$$\Delta z = \frac{2\pi l_0}{b}. \tag{5.69}$$

However, l_0 depends on the index n_0 at the starting point and also on the angle the ray makes at that point with the z-axis. Therefore, all rays leaving a given point within the medium do not have the same period. For paraxial rays, l_0 is approximated by N_0, the axial value of the gradient function, so that these rays do provide a sharp image formation, within the limits of the paraxial approximation. The paraxial image plane is located at a distance of $2\pi N_0/b$ from the object plane or at any multiple of that distance, provided the rod is sufficiently long. Actually, other forms of index profile may also produce the same paraxial imaging.

In some applications it may be unimportant whether the image is inverted or not. Inspection of the paraxial tracing formulas, Eqs. (5.39) and (5.41), shows that a paraxial image plane occurs at the half-period distance $\pi N_0/b$, but with $x = -x_0$, $y = -y_0$. This shows that such a device can be used as an image inverter. Its action would be similar to a fiber-optics "twister" consisting of a bundle of homogeneous fibers arranged in a twisted configuration to invert an image.

Chapter 6
RAY TRACING IN A RADIAL GRADIENT

6.1 Introduction

Most investigations of radial gradients are based on the assumption that the refractive index can be expressed in the form

$$n = N_0 + N_1 r^2 + N_2 r^4 + \cdots, \tag{6.1}$$

where N_0, N_1, N_2, etc., are constants. This is, perhaps, not the most logical form to use since n^2 is a more natural quantity than n itself, as the discussion in Section 5.3 would suggest. Furthermore, it is possible for gradients to exist that would require odd powers of r to be present in Eq. (6.1). However, linear terms in r would make paraxial rays behave badly and so would make conventional aberration theory unworkable.

As pointed out in the last chapter, practical numerical ray tracing in a radial gradient can usually be carried out effectively by using a computer subroutine for solving the differential equations given in Eqs. (5.30). This procedure gives a highly accurate method of determining the location and direction of each ray at each point of its path. The routine applies to both meridional and skew rays. However, other tracing methods have also been developed, and some of these appear preferable for particular purposes.

6.2 Paraxial Ray Trace

Paraxial rays are approximations to the actual rays in an optical system. The paths of these rays are obtained by neglecting second- and higher-order powers (relative to unity) of ray heights and ray slopes in the tracing formulas. These fictitious rays serve as good approximations to the real rays when the rays lie near the optical axis and provide simple approximate predictions of the imaging properties of an optical system. Furthermore, most aberration theories are based on the deviations of the actual rays from the paraxial rays since the paraxial image formation is ideal in many respects.

Since the tracing formulas involve dn/dr, it is necessary to include both the first and second terms of Eq. (6.1) in deriving the paraxial formulas, which, therefore, depend on both N_0 and N_1.

A simple way to derive the paraxial formulas is to consider the special gradient given by Eq. (5.34). In series form this gives

$$n = N_0\left(1 \pm \frac{b^2 r^2}{2N_0^2}\right) \tag{6.2}$$

in the paraxial approximation. Thus the paraxial rays would coincide with those of any gradient for which

$$N_1 = \pm \frac{b^2}{2N_0}. \tag{6.3}$$

However, the tracing equations, as given by Eqs. (5.41), (5.42), and (5.45) are exact for all rays, and therefore are also valid for paraxial rays. Therefore, defining

$$b = |2N_1 N_0|^{1/2}, \qquad \bar{z} = (z - z_0)\frac{b}{l_0}, \tag{6.4}$$

we have, for $N_1 < 0$ and a ray starting at $\bar{z} = 0$,

$$x = x_0 \cos\bar{z} + \frac{p_0}{b}\sin\bar{z},$$

$$\tag{6.5}$$

$$y = y_0 \cos\bar{z} + \frac{q_0}{b}\sin\bar{z},$$

and

$$p = p_0 \cos \bar{z} - bx_0 \sin \bar{z},$$
$$q = q_0 \cos \bar{z} - by_0 \sin \bar{z}, \qquad (6.6)$$
$$l = l_0 = N_0.$$

The last formula here follows from Eqs. (3.45) and (6.2) applied to the initial point of the ray with the paraxial approximation taken into account. Correspondingly, for $N_1 > 0$,

$$x = x_0 \cosh \bar{z} + \frac{p_0}{b} \sinh \bar{z},$$
$$\qquad (6.7)$$
$$y = y_0 \cosh \bar{z} + \frac{q_0}{b} \sinh \bar{z},$$

and

$$p = p_0 \cosh \bar{z} + bx_0 \sinh \bar{z},$$
$$q = q_0 \cosh \bar{z} + by_0 \sinh \bar{z}, \qquad (6.8)$$
$$l = l_0 = N_0.$$

Actually, in the paraxial approximation, we can replace $\bar{z}$ by $k(z - z_0)$.

Inspection of these equations shows that the paraxial rays are not straight lines. In fact, when $N_1 < 0$, the rays are periodic, the period being

$$\Delta z = \frac{2\pi l_0}{b}, \qquad (6.9)$$

as seen with the help of Eqs. (6.4). This fact is important in fiber applications. In practice b is often small, so that the device described by Eqs. (5.46) and (5.47) may be useful.

6.3   Special Trace for Meridional Rays

As might be expected, tracing meridional rays tends to be simpler than tracing skew rays. For example, consider the special gradient described in Eq. (5.34) with the negative sign selected but modified to include a fourth-degree term in r, i.e.,

$$n^2 = N_0^2[1 - (br)^2 + \alpha_2(br)^4]. \tag{6.10}$$

It is then found that the trace for meridional rays can be given exactly in terms of known functions. For, with $c = 0$, $m = n$, the first of Eqs. (5.31) reduces to

$$\dot{r} = \pm\left[\{1 - (br)^2 + \alpha_2(br)^4\}\frac{N_0^2}{l_0^2} - 1\right]^{1/2}, \tag{6.11}$$

giving

$$z = z_0 \pm \frac{l_0}{N_0}\int_{r_0}^{r}\frac{dr}{[1 - (l_0/N_0)^2 - (br)^2 + \alpha_2(br)^4]^{1/2}}. \tag{6.12}$$

It is clear that the integral involved is an elliptic integral, so that r can be expressed in terms of z by means of elliptic functions. We do not pursue this example further here, since the gradient assumed is somewhat specialized and the tracing formula obtained applies only to meridional rays. Also, the algebra required to reduce the integral to a standard form leads to rather intricate formulas. Again, in this example, it appears that n^2 is a more natural parameter than n in the analysis of gradient problems.

6.4   A Perturbation Tracing Method

Closely related to the previous example is an analytical ray-tracing method developed by Paxton and Streifer (1971a). They consider index functions defined by

$$n^2 = N_0^2\left[1 - \delta\left(\frac{r}{r_0}\right)^2 + \alpha_2\delta^2\left(\frac{r}{r_0}\right)^4 + \cdots\right], \tag{6.13}$$

choosing only the case in which the negative sign is present in the second term. Here N_0, r_0, δ, α_2, α_3, etc., are constants that describe the form of the gradient.

If now x, y, and r are scaled by defining

$$\bar{x} = \frac{x}{r_0}, \qquad \bar{y} = \frac{y}{r_0}, \qquad \bar{r} = \frac{r}{r_0}, \qquad (6.14)$$

and z is scaled by

$$\bar{z} = \frac{z\delta^{1/2}N_0}{r_0 l_0}, \qquad (6.15)$$

it is found that Eq. (5.36), in component form, can be written

$$\frac{d^2\bar{x}}{d\bar{z}^2} + \bar{x} = \bar{x} \sum_{j=2}^{\infty} j\alpha_j(\delta\bar{r}^2)^{j-1}, \qquad \frac{d^2\bar{y}}{d\bar{z}^2} + \bar{y} = \bar{y} \sum_{j=2}^{\infty} j\alpha_j(\delta\bar{r}^2)^{j-1}. \qquad (6.16)$$

When all the coefficients α_2, α_3, ... are zero, these equations reduce to Eqs. (5.40) if account is taken of the slightly different scaling used in the earlier section. In that case a simple solution was found.

If, however, any of the coefficients α_j is nonzero, Eqs. (6.16) are coupled and nonlinear and so are difficult to solve, particularly for the case of skew rays. In the reference cited above, algebraic solutions are developed in the form of power series in δ, i.e.,

$$\begin{aligned}
\bar{x} &= a\cos\psi + \delta x_1(a,b,\psi,\phi) + \delta^2 x_2(a,b,\psi,\phi) + \cdots, \\
\bar{y} &= b\cos\phi + \delta y_1(a,b,\psi,\phi) + \delta^2 y_2(a,b,\psi,\phi) + \cdots,
\end{aligned} \qquad (6.17)$$

where

$$\begin{aligned}
\dot{a} &= \delta A_1(a,b,\theta) + \delta^2 A_2(a,b,\theta) + \cdots, \\
\dot{b} &= \delta B_1(a,b,\theta) + \delta^2 B_2(a,b,\theta) + \cdots, \\
\dot{\psi} &= 1 + \delta C_1(a,b,\theta) + \delta^2 C_2(a,b,\theta) + \cdots, \\
\dot{\phi} &= 1 + \delta D_1(a,b,\theta) + \delta^2 D_2(a,b,\theta) + \cdots.
\end{aligned} \qquad (6.18)$$

We shall not attempt here even to define the various symbols involved in these equations or to give the intricate algebraic formulas required for evaluating them. Suffice it to say that this approach can serve as the basis of a computer routine for accurate tracing of rays in a radial gradient. It is of interest to note that the complexity of the algebra used in developing this theory required use of the FORMAC language whereby the computer was programmed to carry out the algebraic derivations.

6.5   Third-Order Trace

In Section 6.2 formulas were derived for tracing rays in the paraxial approximation for a general radial gradient. The next level of approximation is to derive tracing formulas in which quadratic and lower-order terms in ray heights and ray slopes are retained but higher-order terms relative to unity are neglected. The resulting "third-order" tracing formulas can be used to derive the corresponding third-order aberration formulas to be discussed in a later chapter.

We assume an index function of the form

$$n = N_0 + N_1 r^2 + N_2 r^4, \qquad r = (x^2 + y^2)^{1/2}. \tag{6.19}$$

Then, within the third-order theory,

$$n = N_0(1 + \bar{N}_1 r^2), \tag{6.20}$$

$$\frac{dn}{dr} = 2N_1 r(1 + 2\bar{N}_2 r^2), \quad$$

where

$$\bar{N}_1 = \frac{N_1}{N_0}, \qquad \bar{N}_2 = \frac{N_2}{N_1}. \tag{6.21}$$

As a result, Eq. (5.36), in component form, can be written

$$\ddot{x} = \frac{2N_0 N_1 x[1 + r^2(2\bar{N}_2 + \bar{N}_1)]}{l_0^2},$$

$$\ddot{y} = \frac{2N_0 N_1 y[1 + r^2(2\bar{N}_2 + \bar{N}_1)]}{l_0^2}. \tag{6.22}$$

As before, a dot means d/dz.

We shall assume $N_1 < 0$, in which case the solution involves periodic functions. The case $N_1 > 0$ can be developed in a strictly analogous fashion, but involving hyperbolic functions. With $N_1 < 0$ and abbreviations defined by

$$b = |2\bar{N}_1|^{1/2}, \qquad k = \frac{b}{N_0}, \qquad \hat{N} = -(2\bar{N}_2 + \bar{N}_1), \qquad (6.23)$$

we have

$$\ddot{x}\left(\frac{l_0}{b}\right)^2 + x = \hat{N}xr^2, \qquad (6.24)$$

with a similar equation for y. Setting, as in Eq. (6.4),

$$\bar{z} = (z - z_0)\frac{b}{l_0}, \qquad (6.25)$$

we have

$$\frac{d^2x}{d\bar{z}^2} + x = \hat{N}xr^2,$$

$$\frac{d^2y}{d\bar{z}^2} + y = \hat{N}yr^2. \qquad (6.26)$$

These equations are coupled since r depends on both x and y. However, if the right-hand sides are replaced by zero, the paraxial solutions are given by Eqs. (6.5) in the form

$$x = x_0 \cos \bar{z} + \bar{p}_0 \sin \bar{z},$$

$$y = y_0 \cos \bar{z} + \bar{q}_0 \sin \bar{z}, \qquad (6.27)$$

where

$$\bar{p}_0 = \frac{p_0}{b}, \qquad \bar{q}_0 = \frac{q_0}{b}, \qquad (6.28)$$

and

$$p_0 = n_0\left(\frac{dx}{ds}\right)_0, \qquad q_0 = n_0\left(\frac{dy}{ds}\right)_0 \qquad (6.29)$$

are the optical direction cosines at the initial point of the ray ($\bar{z} = 0$).

Within third-order theory it is permissible to use Eqs. (6.27) to evaluate x, y, and r on the right-hand sides of Eqs. (6.26). In this approximation we obtain from Eqs. (6.27)

$$r^2 = u\cos^2\bar{z} + v\cos\bar{z}\sin\bar{z} + w\sin^2\bar{z} \tag{6.30}$$

with

$$u = x_0^2 + y_0^2, \qquad v = 2(x_0\bar{p}_0 + y_0\bar{q}_0), \qquad w = \bar{p}_0^2 + \bar{q}_0^2. \tag{6.31}$$

As a result, Eqs. (6.26) become uncoupled, and each can be solved by itself.

The solution of the first differential equation, if $z_0 = 0$, as shown in Appendix C, is

$$x = cx_0 + s\bar{p}_0 + kzW(sx_0 - c\bar{p}_0)$$
$$+ \frac{\hat{N}}{8}[a_1cs^2 + a_2s^3 + b_1(s - ckz) + b_2skz], \tag{6.32}$$

where

$$c = \cos(kz), \qquad s = \sin(kz), \qquad W = \bar{N}_1(u + w), \tag{6.33}$$

and

$$a_1 = x_0(u - w) - \bar{p}_0v, \qquad b_1 = \bar{p}_0(u + 3w) + x_0v,$$
$$a_2 = \bar{p}_0(u - w) + x_0v, \qquad b_2 = x_0(3u + w) + \bar{p}_0v. \tag{6.34}$$

In the course of the solution, the first optical direction cosine is found to be given by

$$\bar{p} = c\bar{p}_0 - sx_0 + kzW(s\bar{p}_0 + cx_0)$$
$$+ \frac{\hat{N}}{8}[a_1s(3c^2 - 1) + 3a_2cs^2 + b_1skz + b_2(s + ckz)]. \tag{6.35}$$

The corresponding formulas for y and $\bar{q}$ are obtained from Eqs. (6.32), (6.34), and (6.35) by replacing x_0 by y_0 and $\bar{p}_0$ by $\bar{q}_0$. It is evident that these solutions satisfy the initial conditions (when $z = 0$, $s = 0$, $c = 1$). If $z_0 \neq 0$, z should be replaced by $z - z_0$ in the above equations.

Chapter 7
ABERRATION THEORY

7.1 Basic Concepts

Although it is assumed that the reader is familiar with the fundamentals of aberration theory, we shall discuss these briefly for the case of an optical system having rotational symmetry about the z-axis. It is known that, if the real rays obeyed the paraxial ray-tracing formulas, the imaging properties of the system would be ideal in a number of ways. The resulting theory, called Gaussian optics, does give an approximate description of the behavior of the system and also serves as a reference with which the behavior of the real rays can be compared.

In Gaussian optics, all members of a bundle of rays leaving a given point P_0 of the object space meet again at a single image point P' in the image space. In fact, if various points P_0 lie on a given object plane perpendicular to the z-axis, the corresponding image points P' lie on a corresponding plane in the image space. Furthermore, if r_0 and r' give the radial distances of P_0 and P' from the axis, the magnification $m = r'/r_0$ is constant for all points of the given object plane.

These and other features of Gaussian optics remain valid even if some of the media in the system are inhomogeneous, provided, of course, that any gradients present are symmetrical about the z-axis.

The classical approach to aberration theory is to consider an object point P_0 and its Gaussian image point P' and then to consider the real rays starting at P_0 and passing near the point P' in the image space. In particular, if $\Delta x'$ and $\Delta y'$ are the two lateral deviations of the x- and y-coordinates (in the Gaussian image plane) from those at P', then $\Delta x'$ and $\Delta y'$ are measures of the aberrations for the real ray selected.

Since formulas for $\Delta x'$ and $\Delta y'$ in terms of the ray parameters and the construction parameters of the system are complicated, it is customary to expand these two functions into power series. Then one can examine, in their turn, the terms of various degrees. For technical reasons these are usually referred to as terms of third order, fifth order, etc., and the values obtained in this way are called third-order aberrations, fifth-order aberrations, etc.

Each group in the series expansions consists of a sum of terms, each term having a special kind of influence on the imaging qualities of the system. Thus, for example, there are five third-order aberrations: spherical aberration, coma, curvature of field, astigmatism, and distortion.

There are number of advantages to this theory of aberrations. The general effects of the various kinds of aberrations, so defined, are known, and this is helpful in designing systems where special kinds of imaging qualities are desired. Furthermore, each of the individual aberrations can be expressed as the sum of contributions from the various surfaces and media within the system. Thus, in the design process, it is possible to determine which parameters should be adjusted in order to optimize the system. Another advantage of this formulation of aberration theory is that only paraxial rays need to be traced.

A drawback of the classical theory is that the overall analysis becomes very elaborate. For, if the system is optimized relative to the third-order aberrations, the effects of the fifth-order aberrations must still be considered. There are many more of these and the formulas for them are more complicated. Actually, in high-quality systems, it is necessary to calculate both third-order and fifth-order aberrations and to use the aberrations of each order to cancel out those of the other as far as possible. Therefore, lens design remains partly a science and partly an art.

In view of the difficulties inherent in the classical approach to aberration theory, an alternative method can be considered. This is to evaluate the total aberrations directly. This can be done by tracing both real and paraxial rays and computing $\Delta x'$ and $\Delta y'$ for various object points and various rays. Here the tremendous speed of modern computers can be brought to bear. Thousands of rays can be traced quickly, and various combinations of parameters can be tried with a view to optimizing the system.

A common design technique is to plot, for a given object point, the intersection points of a large number of rays with the image plane. The resulting "spot diagram" can then be viewed and judged with respect to quality. In fact, one can define a figure of merit, such as the radius of gyration, of such a spot diagram, so that the computer can give an objective estimate of the image quality.

Methods such as that described above are in common use at the present time. Some of these involve the modulation transfer criterion to give practical information about the ability of a system to image details of various dimensions.

The direct calculation of $\Delta x'$ and $\Delta y'$, while simple to do, is not, however, without its own drawbacks. For there is very little information available to the designer to indicate how to change the many parameters of the system so as to optimize its performance. Consequently, one can consider adopting an approach that is a combination of the two methods outlined above. Motivated by the historical studies of the classical aberrations of various kinds, we can define total aberrations of types analogous to these.

7.2 Total Aberrations

Five total monochromatic aberrations can be defined as suggested by the five classical third-order aberrations. First, however, it is assumed that, by paraxial ray tracing, the location of an image plane has been determined corresponding to the given object plane. If the object plane being considered is located at infinity, the image plane is called the *Gaussian focal plane.*

Spherical aberration is defined as follows. A real ray is traced from the axial point of the object plane to its intersection with the image plane. The distance $\bar{S}$ of this intersection point from the axis is called the *lateral spherical aberration* for this ray. The corresponding distance $\bar{S}$ for a paraxial ray would be zero. If the same real ray is traced to its image-side intersection with the axis at a distance of S from the image plane, S is called the *longitudinal spherical aberration* for this ray. Evidently, S can be either positive or negative.

Meridional coma is defined as follows, where we now must consider the size and position of the entrance pupil present in the system. Three

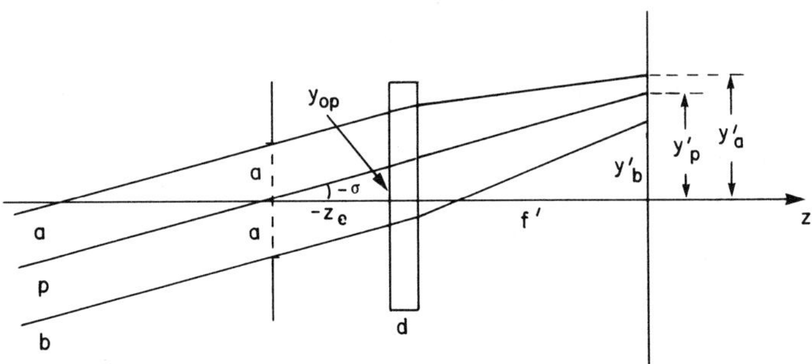

Fig. 7.1. Definition of meridional coma (from Marchand, 1976).

real meridional rays are traced from the object point, the a-ray passing through the top, the b-ray through the bottom, and the p-ray through the center of the entrance pupil. Figure 7.1 illustrates the situation where we have assumed an object point at infinity so that the three rays are parallel in the object space. The location of the "object point" P_0 in such a case is considered to be specified by the direction of the parallel bundle of rays being considered. For simplicity the optical system is pictured in the figure as a plane parallel plate, which would be a Wood lens if it contained a radial gradient.

With the object plane at infinity the image plane coincides with the Gaussian focal plane. In any case, for three such rays in the yz-plane, image heights y'_a, y'_b, and y'_p are obtained. Then the meridional coma C is defined by

$$C = \frac{1}{2}(y'_a + y'_b) - y'_p. \tag{7.1}$$

We shall not discuss in detail the motivation behind this definition other than to assert that the quantity C is closely related to the familiar cometlike pattern produced by an optical system when this aberration is present.

Distortion is defined by considering the p-ray, mentioned above, and comparing its height in the image plane with the height y'_0 of the paraxial image of the given object point. Thus, distortion, in percent, is defined by

$$D = \frac{100(y'_p - y'_0)}{y'_0}. \tag{7.2}$$

The aberrations known as meridional and sagittal curvature have to do with the tendency of a bundle of rays from a given object point to converge not at the Gaussian image plane but at two points located away from that plane. This means that points on a given object plane tend to be imaged, not on the Gaussian plane, but on a curved surface having two unequal curvatures at each point. The difference of these two curvatures is called astigmatism.

The motivation for the definition of meridional curvature is suggested by Fig. 7.2, where again we assume an object point at infinity. Here we consider the same p-ray as before and also a parallel ray in the same meridional plane but shifted upwards by a distance h in the object space. Then the meridional curvature Z_m is defined by

$$Z_m = \lim_{h \to 0} \frac{y' - y'_p}{u' - u'_p}, \tag{7.3}$$

where u' and u'_p are the slopes of the two rays in the image space.

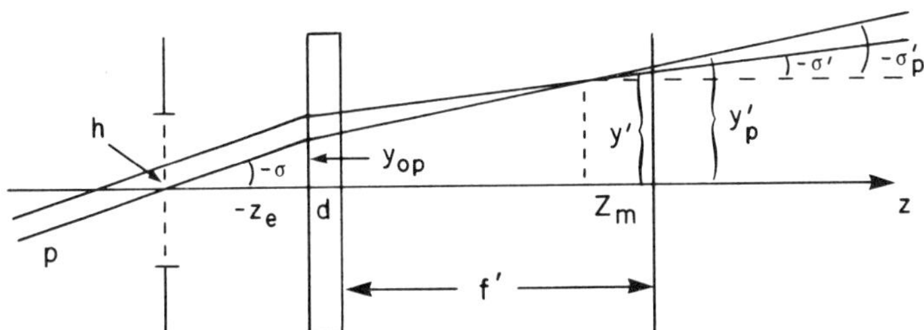

Fig. 7.2.   Definition of meridional curvature (from Marchand, 1976).

The analogous definition for the sagittal curvature Z_s requires the consideration of a skew ray. Pictured in Fig. 7.3 is the same meridional p-ray introduced above and a second ray parallel to it, but shifted a distance h' away from it in the x-direction in the object space. Then, if x' is the x-coordinate of this second ray as it intersects the image plane and p' and l' are the first and third direction cosines of the ray in the image space, we define Z_s by

$$Z_s = \lim_{h' \to 0} \frac{x' l'}{p'}. \tag{7.4}$$

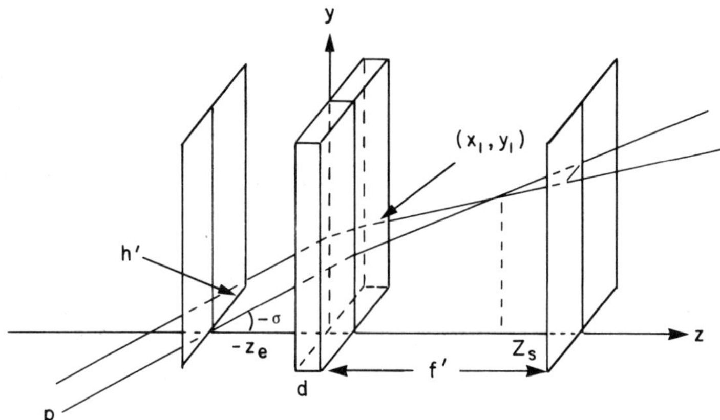

Fig. 7.3. Definition of sagittal curvature (from Marchand, 1976).

Some writers define these two curvatures as the negatives of the above quantities. The astigmatism A is defined by

$$A = \frac{1}{2}(Z_m - Z_s). \tag{7.5}$$

When this quantity is zero, the bundle of rays from the given object point may converge away from the image plane but at least will not converge at two different places.

Instead of attempting to minimize the two curvatures Z_m and Z_s, it is often customary to try to reduce the so-called Petzval curvature defined by

$$P = Z_m - 3Z_s. \tag{7.6}$$

Analytic formulas for this quantity tend to be particularly simple.

With respect to vignetting, which is not a true aberration, it suffices to say that its evaluation in a system containing gradients is carried out in the traditional manner based on paraxial formulas.

The preceding formulas are for the monochromatic aberrations. Their dependence on wavelength (chromatic aberration) can be determined only when reliable dispersion information is at hand. At present not many data of this sort are available for gradient materials.

7.3 Buchdahl Theory

Buchdahl (1969, 1970) developed a systematic method of computing the order-by-order classical aberrations of an optical system having rotational symmetry about an axis. Details of the case for third-order aberrations, with special attention to the presence of gradient elements, have been worked out by Sands (1970, 1971a,b,c,d, 1973).

In the Buchdahl–Sands theory the refractive index of each medium is assumed to be given in the form

$$n = N_0(z) + N_1(z)\xi + N_2(z)\xi^2 + \cdots, \tag{7.7}$$

where

$$\xi = r^2 = x^2 + y^2. \tag{7.8}$$

With n a function of both r and z, a method is required for ray tracing in such a medium. However, only paraxial tracing is required, so that the appropriate formulas are not difficult to obtain.

In a symmetrical optical system of the type considered, the third-order vectorial aberration

$$\varepsilon = (\Delta x', \Delta y') \tag{7.9}$$

can be described by an expression of the form (Buchdahl, 1970)

$$\varepsilon = \left[\sigma_1\xi_1 + 2\sigma_2\eta_1 + (\sigma_3 + \sigma_4)\zeta_1\right]\mathbf{S} \\ + \left[\sigma_2\xi_1 + 2\sigma_3\eta_1 + \sigma_5\zeta_1\right]\mathbf{T}. \tag{7.10}$$

Here

$$\mathbf{S} = (S_x, S_y), \qquad \mathbf{T} = (T_x, T_y) \tag{7.11}$$

are the two-dimensional vectors indicating the intersection point of the ray with the object plane and entrance pupil respectively, and

$$\begin{aligned} \xi_1 &= S_x^2 + S_y^2, \\ \eta_1 &= S_x T_x + S_y T_y, \\ \zeta_1 &= T_x^2 + T_y^2. \end{aligned} \tag{7.12}$$

It is not difficult to verify that the five coefficients σ_1 to σ_5 are the familiar Seidel third-order aberration coefficients. The coefficient of spherical aberration is σ_1, of coma is σ_2, of astigmatism is σ_3, of Petzval curvature is σ_4, and of distortion is σ_5.

In Buchdahl's theory each of these coefficients is expressed as the sum of contributions from all the surfaces and intersurface transitions in the optical system. These contributions can be computed by tracing two meridional paraxial rays through the system. The a-ray is defined by heights of unity at the object plane and zero at the entrance pupil, the b-ray by heights of zero at the object plane and unity at the entrance pupil. Tracing these two rays (by paraxial optics) through the system determines the heights $y_a(z)$, $y_b(z)$ and the slopes $v_a(z) = \dot{y}_a(z)$, $v_b(z) = \dot{y}_b(z)$ for all points. The dot indicates, as before, differentiation with respect to z.

For a typical refracting surface having curvature c, Sands defines a constant K defined by

$$K = -c\,\Delta\left(2N_1 + \frac{c\dot{N}_0}{2}\right),\tag{7.13}$$

Δ indicating the change in a quantity in moving from one side of a surface to the other. Here N_0 and N_1 are the coefficients in Eq. (7.7) for either of the two media involved. If both media are homogeneous, $K = 0$.

By way of further notation, let

$$k_0 = \frac{N_0}{N_0'}\tag{7.14}$$

represent the ratio of the axial refractive indices and

$$i_a = v_a + cy_a,$$
$$i_b = v_b + cy_b,$$
$$q = \frac{i_b}{i_a},\tag{7.15}$$
$$\lambda = N_0(z)\left[\,y_a(z)v_b(z) - y_b(z)v_a(z)\right],$$
$$a = \frac{1}{2}\,N_0(k_0 - 1)y_a i_a^2(i_a + v_a').$$

Evidently i_a is the angle of incidence, i.e., the angle the a-ray makes with the surface normal before refraction, and i_b correspondingly for the b-ray. The quantity λ is a paraxial invariant and so can be determined at any convenient point of the ray. The quantity a is the usual contribution of a spherical surface to the spherical aberration.

With the above notation it is found that the contributions of the surface to the five coefficients σ_1 to σ_5 are given by

$$
\begin{aligned}
a_1 &= a + K y_a^4, \\
a_2 &= aq + K y_a^3 y_b, \\
a_3 &= aq^2 + K y_a^2 y_b^2, \\
a_4 &= \frac{1}{2} \lambda^2 c \, \Delta\!\left(\frac{1}{N_0}\right), \\
a_5 &= aq^3 + qa_4 + K y_a y_b^3.
\end{aligned}
\tag{7.16}
$$

The effect of a transfer from any surface to the next one is obtained from the contributions a_1^* to a_5^* given by

$$
\begin{aligned}
a_1^* ={}& \frac{1}{2} \nabla(N_0 y_a v_a^3) + \int \left[4N_2 y_a^4 + 2N_1 y_a^2 v_a^2 - \frac{1}{2} N_0 v_a^4 \right] dz, \\[1em]
a_2^* ={}& \frac{1}{2} \nabla(N_0 y_a v_a^2 v_b) \\
&+ \int \left[4N_2 y_a^3 y_b + N_1 y_a v_a(y_a v_b + y_b v_a) - \frac{1}{2} N_0 v_a^3 v_b \right] dz, \\[1em]
a_3^* ={}& \frac{1}{2} \nabla(N_0 y_a v_a v_b^2) \\
&+ \int \left[4N_2 y_a^2 y_b^2 + 2N_1 y_a y_b v_a v_b - \frac{1}{2} N_0 v_a^2 v_b^2 \right] dz, \\[1em]
a_4^* ={}& \lambda^2 \int (N_1/N_0^2)\, dz, \\[1em]
a_5^* ={}& \frac{1}{2} \nabla(N_0 y_a v_b^3) \\
&+ \int \left[4N_2 y_a y_b^3 + N_1 y_b v_b(y_a v_b + y_b v_a) - \frac{1}{2} N_0 v_a v_b^3 \right] dz.
\end{aligned}
\tag{7.17}
$$

Here the symbol ∇ indicates the change of the following quantities in the transfer from one surface to the next.

With the above formulas it is seen that the computation of the Seidel (third-order) aberration coefficients is relatively simple for a system having rotational symmetry even if several of the elements are inhomogeneous. The five coefficients are given by

$$\sigma_i = \mu\left(\sum_j a_{ij} + \sum_j a^*_{ij}\right), \tag{7.18}$$

where

$$\frac{1}{\mu} = N_0(z_0)v_a(z'), \tag{7.19}$$

z_0 and z' being the z-coordinates of the object and image planes respectively. The two summation signs in Eq. (7.18) indicate that each of the coefficients σ_i is obtained by summing the surface contributions a_i over all surfaces and the transfer contributions a^*_i over all the intervals in the system.

In the important case when n is independent of z (radial gradient), the ray tracing can be done by Eqs. (6.5) and (6.6) when $N_1 < 0$, and by Eqs. (6.7) and (6.8) when $N_1 > 0$. In this case $N_0, N_1, N_2, \ldots$ are constant and the required integrals have the form

$$\int y_a^4\, dz, \qquad \int y_a^2 v_a^2\, dz, \qquad \int v_a^4\, dz, \qquad \int y_a v_a^2 v_b\, dz, \tag{7.20}$$

etc. But, for a radial gradient, the paraxial quantities y_a, y_b, v_a, v_b are given by simple functions involving $\sin \bar{z}$ and $\cos \bar{z}$ when $N_1 < 0$, and $\sinh \bar{z}$ and $\cosh \bar{z}$ when $N_1 > 0$. Consequently, these integrals can be evaluated once and for all in the indefinite form. Then it remains only to insert the appropriate z values as limits of integration, i.e., the axial values of z at the surfaces involved.

From Eqs. (6.5) we have, if $z_0 = 0$,

$$y_a = y_{0a} \cos(kz) + \frac{q_0}{b} \sin(kz), \tag{7.21}$$

in the paraxial approximation, for a medium with $N_1 < 0$. For a meridional ray we have

$$q = -n \sin \sigma \tag{7.22}$$

where σ is the inclination angle of the ray, σ being positive or negative as the ray is falling or rising as z increases (the usual optical convention). In paraxial optics

$$v = -\tan\sigma = \frac{q}{N_0} \tag{7.23}$$

since the tangent can be replaced by the sine and n by its axial value. Hence, from Eqs. (6.6),

$$v_a = v_{0a}\cos(kz) - ky_{0a}\sin(kz). \tag{7.24}$$

The formulas for y_b and v_b are analogous.

Substitution of the formulas for y_a, y_b, v_a, and v_b into Eq. (7.20) shows that several elementary integrals must be evaluated, namely,

$$\int \cos^4(kz)\,dz, \qquad \int \cos^3(kz)\sin(kz)\,dz,$$

$$\int \cos^2(kz)\sin^2(kz)\,dz, \tag{7.25}$$

$$\int \cos(kz)\sin^3(kz)\,dz, \qquad \int \sin^4(kz)\,dz.$$

As already pointed out, these integrals can be evaluated once and for all in closed form, so that no numerical quadratures are required. The case with $N_1 > 0$ is entirely analogous, involving hyperbolic instead of trigonometric functions.

The attractive technique outlined above has many advantages. However, it must be kept in mind that the formulas given here apply only to the third-order aberrations. These are not adequate for the design of high-quality optical systems. Furthermore, making the third-order aberrations small may not even be desirable, so that minimizing these aberrations may not be a useful start in the design of a system.

Recently the Buchdahl theory has been extended to the fifth-order aberrations (Gupta *et al.*, 1976). This is an important step forward. We shall not give the explicit formulas since they are quite extensive. However, it is well known that the fifth-order and higher values do constitute significant contributions to the total aberrations in practical optical systems.

Chapter 8

THE WOOD LENS

8.1 Introduction

The Wood lens described in Chapter 1 consists of a plane parallel plate with a radial gradient, the axis of symmetry being perpendicular to the faces. This is one of the simplest possible optical systems involving a gradient, and serves as a good example to illustrate the methods used to deal with gradient index lenses.

A Wood lens can be designed with surprisingly good imaging properties considering that no curved surfaces are required. This emphasizes the fact that gradient index media provide lens designers with several new degrees of freedom in accomplishing various design objectives.

Actually, a Wood lens differs only slightly from a radial gradient fiber, since these usually have plane faces at the ends. The chief difference is that the fiber is usually long and narrow, whereas the Wood lens is usually short with a sizable radius. Although mathematically the two appear to be much the same, the periodic nature of rays in the fiber is of great importance but plays no real role in the Wood lens. Also fibers with $N_1 > 0$ appear to be of little practical use. What is more, in fiber optics the imaging qualities within the fiber are of prime interest, whereas with a Wood lens the object and image planes are usually entirely outside the lens.

8.2 The Photographic Wood Lens

As an illustrative example we consider a Wood lens with object plane at infinity. For this case, the ray-tracing formulas are particularly simple since refraction at a plane surface involves only a trivial calculation. Of course, the third-order aberration formulas in the Buchdahl theory are relatively simple for the same reason. However, instead of following this approach, we shall make use of the third-order ray-tracing formulas introduced in Chapter 6 for the case $N_1 < 0$. These formulas combined with the aberration definitions given in Chapter 7 provide a means of deriving simple formulas for the third-order aberrations of a photographic Wood lens (see Marchand, 1976).

In considering a meridional ray, it is seen that the quantity

$$q = -n \sin \sigma \tag{8.1}$$

is unchanged in refraction at a plane surface (by Snell's law). With this in mind, the paraxial information for the lens can be found from Eqs. (6.5) and (6.6) for the case with $N_1 < 0$. For a lens thickness of d, we write

$$k = |2\bar{N}_1|^{1/2}, \qquad \bar{k} = kd,$$
$$c = \sin \bar{k}, \qquad s = \cos \bar{k}, \tag{8.2}$$

giving

$$x_1 = cx_0 + s\bar{p}_0,$$
$$y_1 = cy_0 + s\bar{q}_0, \tag{8.3}$$

and

$$\bar{p}_1 = c\bar{p}_0 - sx_0,$$
$$\bar{q}_1 = c\bar{q}_0 - sy_0 \tag{8.4}$$

for the paraxial trace within the lens up to the second surface. Here $\bar{p}_0 = p_0/b$, etc., with $b = N_0 k$.

To find the focal length we consider an object ray in the plane $x = 0$, parallel to the axis, a distance h away from it, and with its image ray intersecting the axis at an angle σ'. Then f is defined by

$$f = \lim_{h \to 0} \frac{h}{\tan \sigma'}. \tag{8.5}$$

From the last of Eqs. (8.4) with $q_0 = 0$ and $y_0 = h$, we find that

$$q_1 = b\bar{q}_1 = -hbs = q' = -\sin \sigma' = -\tan \sigma' \tag{8.6}$$

in the paraxial approximation, where we also noted that q_1 is unchanged in refraction at the second surface. Hence, from Eq. (8.5),

$$f = \frac{1}{bs} = \frac{1}{b \sin(kd)}. \tag{8.7}$$

It is interesting to note that, for the case $N_1 > 0$, Eq. (6.8) leads to the formula

$$f = \frac{-1}{b \sinh(kd)}. \tag{8.8}$$

This shows that a Wood lens acts like a converging or diverging lens depending on whether $N_1 \lessgtr 0$. If kd is small (a thin low-power lens), series development of Eqs. (8.7) and (8.8) shows that

$$f = \frac{-1}{2N_1 d} \tag{8.9}$$

in both cases. This formula was given by Wood about 1900.

The back-focus f' (distance of the focal point from the second surface) is defined by

$$f' = \lim_{h \to 0} \frac{y_1}{\tan \sigma'} \tag{8.10}$$

applied to the same ray as before. For the case $N_1 < 0$, Eq. (8.6) and the second of Eqs. (8.3) with $q_0 = 0$, $y_0 = h$, lead to

$$f' = \frac{c}{bs} = cf. \tag{8.11}$$

With the object plane at infinity the image plane coincides with the Gaussian focal plane.

For a given object point (here assumed to lie in the plane $x = 0$ and specified by a value of its inclination angle σ in the object space) the paraxial image height y'_0 can be found as follows. The value of

q_0 just after the first surface is given by

$$q_0 = -n_0 \sin \sigma_0 = -\sin \sigma \qquad (8.12)$$

since it is invariant across the first surface and the index in air is unity. By simple geometry it is clear that

$$y_1 - y_0' = f' \tan \sigma' \qquad (8.13)$$

or

$$y_0' = y_1 + f' b \bar{q}_1 \qquad (8.14)$$

in the paraxial approximation, where again Eq. (8.6) was noted.

The paraxial values of y_1 and q_1 are given by Eqs. (8.3) and (8.4). Thus

$$y_1 = c y_0 + s \bar{q}_0, \\ \bar{q}_1 = c \bar{q}_0 - s y_0, \qquad (8.15)$$

with c and s defined by Eqs. (8.2). Substitution of these together with Eq. (8.11) into Eq. (8.14) leads to

$$y_0' = \frac{\bar{q}_0}{s} = f q_0. \qquad (8.16)$$

The third-order aberrations of the photographic Wood lens can be obtained by applying the third-order ray-tracing formulas given in Chapter 6 to the aberration definitions presented in Chapter 7. For a ray in the yz-plane we have by analogy with Eqs. (6.32) to (6.35)

$$y_1 = c y_0 + s \bar{q}_0 + \bar{k} W (s y_0 - c \bar{q}_0) \\ + \frac{\hat{N}}{8} \left[a_1 c s^2 + a_2 s^3 + b_1 (s - c \bar{k}) + b_2 s \bar{k} \right] \qquad (8.17)$$

and

$$\bar{q}_1 = c \bar{q}_0 - s y_0 + \bar{k} W (s \bar{q}_0 + c y_0) \\ + \frac{\hat{N}}{8} \left[a_1 s(3c^2 - 1) + 3 a_2 c s^2 + b_1 s \bar{k} + b_2 (s + c \bar{k}) \right], \qquad (8.18)$$

where now

$$\bar{k} = kd, \qquad c = \cos\bar{k}, \qquad s = \sin\bar{k}, \tag{8.19}$$

and

$$\begin{aligned}
a_1 &= y_0(u - w) - \bar{q}_0 v, & b_1 &= \bar{q}_0(u + 3w) + y_0 v, \\
a_2 &= \bar{q}_0(u - w) + y_0 v, & b_2 &= y_0(3u + w) + \bar{q}_0 v.
\end{aligned} \tag{8.20}$$

The quantities u, v, w, and W are defined the same as before, as in Eqs. (6.31) and (6.33).

The five third-order aberrations of the photographic Wood lens can be derived by using the third-order tracing formulas.

The intersection height y' of a meridional ray (in the yz-plane) at the focal plane is given by

$$\frac{y_1 - y'}{f'} = \tan\sigma' \tag{8.21}$$

or

$$y' = y_1 + f'b\bar{q}_1. \tag{8.22}$$

After substituting Eqs. (8.17) and (8.18) into this equation and carrying out some tedious but straightforward algebra, we find

$$\begin{aligned}
y' = {}& q_0 f + cN y_0(u + w) \\
& + \frac{1}{2} c^3 q_0 f' \left[q_0^2 + q_0 \frac{y_0}{f'}(t^2 - 3) + 3\left(\frac{y_0}{f'}\right)^2 \right] \\
& + \hat{N} c^3 \left\{ \frac{t\bar{q}_0}{4}\left[3u(2 + t^2) + wt^2 \right] - y_0 w \right\},
\end{aligned} \tag{8.23}$$

where

$$t = s/c,$$

$$N = \frac{\bar{N}_1 \bar{k}}{cs} - \frac{1}{2f^2} + \frac{\hat{N}}{8}\left[2c^2 + 3\left(1 + \frac{\bar{k}}{cs}\right) \right]. \tag{8.24}$$

Here the ray being traced is specified by the values of y_0 and $q_0 = -\sin\sigma$.

The lateral spherical aberration $\bar{S}$, as defined in Chapter 7, is found by inspection from Eq. (8.23) by setting $y_0 = h$, $q_0 = 0$, and, from Eqs. (6.31), $u = h^2$, $v = 0$, and $w = 0$. The result is

$$\bar{S} = h^3 cN, \tag{8.25}$$

within the limitations of the third-order theory. The longitudinal spherical aberration S can be deduced from this result by considering the intersection of the same ray with the axis. We have

$$S = \frac{\bar{S}}{\tan \sigma'} = \frac{-\bar{S}}{b\bar{q}_1}, \tag{8.26}$$

since here it is only necessary to use the paraxial approximation for $\tan \sigma'$. Hence, from Eq. (8.15) with $q_0 = 0$ and $y_0 = h$, we find

$$S = h^2 f' N \tag{8.27}$$

for the third-order longitudinal spherical aberration.

The definition of meridional coma was illustrated in Fig. 7.1, where a Wood lens is shown together with three parallel rays from an object point at infinity. The location of the entrance pupil is specified by its distance z_e from the first surface, with $z_e \lessgtr 0$ as the pupil is to the left or right of the surface.

The p-ray passes through the center of the entrance pupil and the a-ray and b-ray through the top and bottom of the entrance pupil, respectively. The ray heights at the first surface are given, within the third-order approximation, by

$$y_{0p} = z_e \tan \sigma = -z_e q_0 \left(1 + \frac{q_0^2}{2} \right),$$

$$y_{0a} = y_{0p} + a, \qquad y_{0b} = y_{0p} - a, \tag{8.28}$$

a being the radius of the pupil. The value of q_0 is the same for all three rays since they are parallel in the object space.

Application of Eq. (8.23) to each of these three rays and substitution into Eq. (7.1) lead, after some algebra, to the result

$$C = \frac{3q_0 a^2 c^2}{2f} \left[1 - 2Nf^2 \hat{z} + \frac{\hat{N}}{2b^2} \left(1 + \frac{1}{c^2} \right) \right], \tag{8.29}$$

where

$$\hat{z} = \frac{z_e}{f'}. \tag{8.30}$$

Equation (8.29) gives the third-order meridional coma for a convergent Wood lens in air with object plane at infinity. It may be noted that, if spherical aberration is corrected, i.e., $N = 0$, then C does not depend on the position of the entrance pupil.

The distortion can be found by considering the p-ray defined above and using Eqs. (8.23) and the definition given by Eq. (7.2). The result is

$$D = 50q_0^2 c^4 \left\{ 1 + \hat{z}(3 - t^2) + 3\hat{z}^2 - 2Nf^2\hat{z}(t^2 + \hat{z}^2) \right.$$
$$\left. + \frac{\hat{N}}{2b^2} [3\hat{z}^2(2 + t^2) + 4\hat{z} + t^4] \right\}. \tag{8.31}$$

The meridional curvature was defined by Eq. (7.3) with reference to Fig. 7.2. In this equation the denominator can be evaluated paraxially, giving

$$u' - u'_p = q_1 - q_{1p} = bsh = \frac{-h}{f}, \tag{8.32}$$

so that Eq. (7.3) can be written

$$Z_m = -f \lim_{h \to 0} \frac{y' - y'_p}{h}. \tag{8.33}$$

Applying Eq. (8.23) to each of the two rays involved and later using the paraxial relation $y_{0p} = -z_e q_0$, we are led to the result

$$Z_m = \left(\frac{1}{2} f' q_0^2 c^2 \right) \left\{ 3(1 + 2\hat{z}) - t^2 - 2Nf^2(t^2 + 3\hat{z}^2) \right.$$
$$\left. + \frac{\hat{N}}{b^2} [2 + 3\hat{z}(2 + t^2)] \right\}. \tag{8.34}$$

for the meridional curvature.

The sagittal curvature, defined by Eq. (7.4) with reference to Fig. 7.3, involves the trace of a skew ray. For the Wood lens case, it is found by geometric arguments that Z_s can be written

$$Z_s = f' + \lim_{h' \to 0} \frac{x_1 l'}{p_1}, \tag{8.35}$$

which involves evaluation of x_1, p_1, and l' for the skew ray. Here

$$l' = (1 - p_1^2 - q_1^2)^{1/2} \tag{8.36}$$

because both p and q are unchanged in refraction at a plane surface. The quantities x_1 and p_1 can be evaluated by analogy with Eqs. (8.17) to (8.20), i.e., by replacing y_0 and q_0 by x_0 and p_0 respectively.

Carrying out the algebra involved and using again the paraxial relation $y_0 = -z_e q_0$, we eventually find

$$Z_s = \left(\frac{1}{2} f' q_0^2 c^2\right)\left\{ 1 + 2\hat{z} - t^2 - 2Nf^2(t^2 + \hat{z}^2) \right.$$
$$\left. + \frac{\hat{N}}{b^2}\left[1 + \frac{1 + \bar{k}/(cs)}{2c^2} + (2 + t^2)\hat{z} \right] \right\} \tag{8.37}$$

for the sagittal curvature.

It is now a simple matter to obtain the astigmatism. In accordance with the definition stated in Eq. (7.5), we take half the difference of Z_m and Z_s, giving

$$A = \left(\frac{1}{2} f' q_0^2 c^2\right)\left\{ 1 + 2\hat{z} - 2Nf^2\hat{z}^2 + \frac{\hat{N}}{2b^2}\left[1 + 2\hat{z}(2 + t^2) \right.\right.$$
$$\left.\left. - \frac{1 + \bar{k}/(cs)}{2c^2} \right]\right\}. \tag{8.38}$$

The Petzval curvature is defined by Eq. (7.6), resulting in the formula

$$P = f' q_0^2 c^2 \left\{ t^2(1 + 2Nf^2) \right.$$
$$\left. - \frac{\hat{N}}{2b^2}\left[1 + \frac{3}{2c^2}\left(1 + \frac{\bar{k}}{cs}\right) \right] \right\}. \tag{8.39}$$

It is remarkable that the third-order aberrations can be expressed by such simple formulas. This is because of the two plane surfaces and also the object plane at infinity. With a finite object distance the corresponding formulas become quite complicated and with curved surfaces much more so. However, the present example gives insight into the nature of the aberrations.

With the above formulas it is a simple matter to determine how to correct or at least optimize the third-order aberrations. As seen from Eqs. (8.24) and (8.25), spherical aberration can be corrected by taking $N = 0$.

Assume that N_0, N_1, and d are chosen to produce a specified focal length f through Eq. (8.7). Then N will vanish if

$$\frac{\hat{N}}{8} = \frac{1/(2f^2) - \bar{N}_1 \bar{k}/(cs)}{2c^2 + 3(1 + \bar{k}/(cs))}, \tag{8.40}$$

which can be satisfied by a suitable choice of N_2 (see Eqs. (6.21) and (6.23)) With $N = 0$ each of the other aberration formulas is simplified. As before, $\bar{k}$, c, and s are given by Eqs. (8.2).

A second aberration can be completely corrected by a suitable choice of the pupil position specified by $\hat{z}$. For example, the astigmatism given by Eq. (8.38) can be corrected, in case $N = 0$, by solving a linear equation for $\hat{z}$.

To illustrate the above correction scheme a numerical example was calculated as described in Tables 8.1 and 8.2, where the dimensions could be in centimeter units. As a check the values were also computed by the Buchdahl–Sands method.

TABLE 8.1

Wood Lens Data[a]

Lens radius	1
Thickness	$d = 0.5$
Gradient index	$N_0 = 1.5$
	$N_1 = -0.05$
	$N_2 = 4.036 \times 10^{-5}$
Focal length	$f = 20.0557$
Back-focus	$f' = 19.8888$
Aperture	f/10
Half-field	$|\sigma| = 14°$
Entrance pupil	$z_e = -8.031$, a = 1
Object plane	at infinity
Maximum image height (from axis)	$y_0' = 5.00$

[a] From Marchand (1976).

TABLE 8.2

Maximum Third-Order Aberrations[a]

Spherical aberration	$S = 0$
Coma	$C = 0.022$
Distortion	$D = 0.6\%$
Petzval curvature	$P = -0.523$
Astigmatism	$A = 0$

[a]From Marchand (1976).

The GRIN rod and Selfoc fibers are essentially the same as Wood lenses, being long and narrow cylinders with plane surfaces at the ends and radial gradients internally. However, the algebraic aberration formulas given above would have to be modified for them since the object distance is not infinity. Ordinarily both the object plane and image plane are considered to be on the ends of or interior to the rod.

The fiber image-relay problem has been treated by Paxton and Streifer (1971b), who applied the ray-tracing methods previously developed by them and used the Sands formulation of the third-order aberration theory.

8.3   The Thin Wood Lens

When d is small, the aberration formulas can be simplified by neglecting quadratic terms in d relative to unity. Thus, a set of thin-lens formulas are obtained that give useful approximate values of the aberrations for many practical cases. In practice

$$|\bar{N}_1| = \left|\frac{N_1}{N_0}\right| < 1, \tag{8.41}$$

so that quadratic terms in $\bar{k} = kd$ can also be neglected relative to unity.

If these approximations are made in the preceding example of a converging Wood lens ($N_1 < 0$) with object at infinity, the following formulas result. The spherical aberration formulas, Eqs. (8.25) and (8.27), reduce to

$$\bar{S} = h^3 N, \qquad S = h^2 f N, \tag{8.42}$$

where

$$N = -\left[2\bar{N}_2 + \frac{1}{2f^2}\right] \tag{8.43}$$

and

$$f = f' = -\frac{1}{2N_1 d}. \tag{8.44}$$

From these relations it follows that spherical aberration can be corrected by choosing N_2 so that $N = 0$, i.e., with

$$N_2 = -N_1^3 d^2. \tag{8.45}$$

The coma formula, Eq. (8.29), reduces to

$$C = \frac{3}{2}\frac{q_0 a^2}{f}\left(1 + \frac{\hat{N}}{b^2} - 2Nf^2\hat{z}\right) \tag{8.46}$$

with

$$\hat{z} = \frac{z_e}{f}, \qquad \hat{N} = -(2\bar{N}_2 + \bar{N}_1). \tag{8.47}$$

Equation (8.31) for distortion takes the form

$$D = 50q_0^2\left[1 + 3\hat{z} + 3\hat{z}^2 - 2Nf^2\hat{z}^3 \right.$$

$$\left. + \frac{\hat{N}}{b^2}(2\hat{z} + 3\hat{z}^2)\right]. \tag{8.48}$$

From Eq. (8.38) for astigmatism we have

$$A = \frac{1}{2}fq_0^2[1 + 2\hat{z}(1 + \hat{N}/b^2) - 2Nf^2\hat{z}^2]. \tag{8.49}$$

In case spherical aberration is corrected ($N = 0$), the astigmatism can also be corrected simply by taking

$$\hat{z} = -\frac{(1/2)}{1 + \hat{N}/b^2}. \tag{8.50}$$

In the thin-lens case the Petzval curvature is given by

$$P = fq_0^2\left[\bar{k}^2(1 + 2Nf^2) - \frac{2\hat{N}}{b^2}\right].\tag{8.51}$$

The simple thin-lens formulas serve as a useful check on the exact third-order aberration formulas.

In the numerical example described in Tables 8.1 and 8.2, we find with the thin-lens formulas

$$\begin{aligned}
f &= f' = 20, \\
N_2 &= 3.125 \times 10^{-5}, \qquad S = 0, \\
\hat{N} &= 0.03458, \\
C &= 0.022, \\
z_e &= -8.966, \qquad A = 0, \\
D &= 0.6\%, \\
P &= -0.520.
\end{aligned}\tag{8.52}$$

As expected, these values agree reasonably well with the exact ones given in Tables 8.1 and 8.2.

With respect to chromatic aberration not much can be said at the present time since not much reliable dispersion information on gradient materials is available. However, a few measured curves show the plots of index versus radial distance as giving nearly parallel curves for two different wavelengths. In this case N_0 changes with wavelength, but N_1, which is proportional to the slope at each point, is independent of λ. As a result, in a thin lens, Eq. (8.44) indicates that the focal length would be independent of λ since it depends on N_1 but not on N_0. This conclusion, however, clearly applies only to the thin Wood lens and only to the extent that the index curves are parallel.

An interesting form of achromatic system was patented by Matsumura *et al.* (1972). It consists of two thick Wood lenses with parameters suitably chosen so as to correct the chromatic aberration.

Chapter 9
MORE GENERAL MEDIA

9.1 Rotation-Symmetric Gradient

The three kinds of gradients considered in earlier chapters (spherical, radial, and axial) are special cases of a medium for which

$$n = f(r, z), \tag{9.1}$$

where n is the distance from the z-axis. In such a gradient the refractive index in any plane $z = $ const. is a function only of the radial distance r. Analysis of this more general type of medium reveals a number of laws and principles that hold for all three special types discussed earlier and some others as well.

The development at the beginning of Chapter 5 proceeds in the same way in the present case for the θ formulas, Eqs. (5.8) and (5.9), involving the skewness invariant c. But with n and m functions of both r and z, Eq. (5.11) must be replaced by

$$\frac{d}{dz}\left[m\dot{r}(1 + \dot{r}^2)^{-1/2}\right] = (1 + \dot{r}^2)^{1/2}\frac{\partial m}{\partial r}. \tag{9.2}$$

If now we let

$$u = m\dot{r}(1 + \dot{r}^2)^{-1/2}, \tag{9.3}$$

we find that

$$\dot{r} = u(m^2 - u^2)^{-1/2} \tag{9.4}$$

and

$$\dot{u} = \left(m\frac{\partial m}{\partial r}\right)(m^2 - u^2)^{-1/2}. \tag{9.5}$$

Also, from Eq. (5.8),

$$\dot{\theta} = \frac{c}{r^2}(m^2 - u^2)^{-1/2}. \tag{9.6}$$

The above three differential equations, with

$$m^2 = n^2 - \frac{c^2}{r^2}, \tag{9.7}$$

define the paths of the rays if suitable initial conditions are specified and n is a known function of r and z. In fact, a standard computer subroutine can be used to solve numerically for r, u, θ, $\dot{r}$, $\dot{u}$, and $\dot{\theta}$ in terms of z.

In general l is not invariant as in the case of a radial gradient. However, from Eq. (9.3) and the first part of Eq. (5.16), it follows that

$$l = (m^2 - u^2)^{1/2}. \tag{9.8}$$

Thus, the above differential equations can be written

$$\dot{r} = \frac{u}{l}, \qquad \dot{u} = \frac{(m\,\partial m/\partial r)}{l}, \qquad \dot{\theta} = \frac{c}{lr^2}, \tag{9.9}$$

with l given by Eq. (9.8). It must be kept in mind, in using these equations, that m depends not only on r and z but also the ray being traced since it depends on c. Once these equations have been solved for r and θ in terms of z, we have

$$x = r\cos\theta, \qquad y = r\sin\theta. \tag{9.10}$$

Since

$$p = nx' = l\dot{x} = l(\dot{r}\cos\theta - r\sin\theta\,\dot{\theta}) \tag{9.11}$$

with a similar equation for q, we find, with the help of Eqs. (9.9.),

$$p = u \cos \theta - \frac{c}{r} \sin \theta$$

$$q = u \sin \theta + \frac{c}{r} \cos \theta.$$

$$(9.12)$$

As usual, the skewness invariant provides the relation

$$xq - yp = x_0 q_0 - y_0 p_0 = c, \tag{9.13}$$

and also the formula

$$p^2 + q^2 + l^2 = n^2 \tag{9.14}$$

holds at every point of a ray.

The aberration theories discussed in Chapter 7 apply to any rotation-symmetric system. However, in the Buchdahl–Sands theory, the integrals in Eqs. (7.17) cannot be evaluated once and for all when the coefficients N_0, N_1, and N_2 depend on z. In practice it is ordinarily necessary to resort to numerical quadratures to evaluate these integrals.

Sands (1971a) has developed a theory of chromatic aberrations for rotation-symmetric systems. His approach is closely related to that described in his earlier paper (Sands, 1970), which in turn rests on Buchdahl's theory of quasi-invariants. Sands' treatment also makes use of Buchdahl's theory of dispersion. We shall not give the details of this work since very little reliable information on the dispersion properties of gradient materials is available at the present time.

9.2 A Practical Ray-Tracing Routine

Moore (1975) has developed another ray-tracing scheme for rotation-symmetric media. He assumes that the refractive index can be approximated by a series of the form

$$n = \sum_{j=0}^{J} \sum_{k=0}^{K} N_{jk} z^j \zeta^k, \tag{9.15}$$

where

$$\zeta = x^2 + y^2. \tag{9.16}$$

He then seeks a solution of the basic differential equation, Eq. (1.7), in polynomial form, i.e.,

$$x = \sum_{i=1} A_i z^{i-1}, \qquad y = \sum_{i=1} B_i z^{i-1}. \tag{9.17}$$

This leads to somewhat intricate formulas for determining the coefficients A_i and B_i when the constants N_{jk} are given. However, once a computer program has been written, this no longer presents a problem.

As usual with power series expansions, the question of convergence arises. If long ray paths are involved, clearly the series for x and y must be carried out to include high powers of z, and this is surely the case if this method is applied to a gradient fiber. Or a succession of dummy surfaces can be introduced so that the series method can be applied successively to each of the spaces between them. In lens elements of moderate thickness the problem of slow convergence is not expected to be troublesome.

9.3   Ray Trace in a General Medium

In the case of a general inhomogeneous medium for which n is a given function of x, y, and z, it is possible to use a computer program for direct solution of the vectorial differential equation (1.7), i.e.,

$$(n\mathbf{r}')' = \nabla n, \tag{9.18}$$

where the prime means differentiation with respect to the arc length s. In component form, of course, there are three second-order equations to be solved.

A convenient way of preparing this computation was suggested by Montagnino (1968). Consider the Taylor series expansion of $\mathbf{r}$ in terms of the arc length s in the neighborhood of a starting point P_0, i.e.,

$$\mathbf{r}(s) = \mathbf{r}(s_0) + \mathbf{r}'(s_0)\,\Delta s + (1/2)\mathbf{r}''(s_0)(\Delta s)^2 + \cdots \tag{9.19}$$

and the corresponding expansion of the derivative

$$\mathbf{r}'(s) = \mathbf{r}'(s_0) + \mathbf{r}''(s_0)\,\Delta s + \cdots. \tag{9.20}$$

As usual in differential geometry

$$\mathbf{t} = \mathbf{r}', \qquad \mathbf{K} = \mathbf{r}'' = \mathbf{t}', \tag{9.21}$$

where $\mathbf{t}$ is the unit tangential vector and $\mathbf{K}$ is the curvature vector of the ray path. Thus Eq. (9.20) can be written

$$\mathbf{t}(s) = \mathbf{t}(s_0) + \mathbf{K}(s_0)\,\Delta s + \cdots. \tag{9.22}$$

To obtain a formula for $\mathbf{K}$, we write Eq. (9.18) in the form

$$n'\mathbf{t} + n\mathbf{K} = \nabla n. \tag{9.23}$$

Since

$$\mathbf{t}^2 = 1, \qquad \mathbf{t}\cdot\mathbf{t}' = \mathbf{t}\cdot\mathbf{K} = 0, \tag{9.24}$$

scalar multiplication of Eq. (9.23) by $\mathbf{t}$ gives

$$n' = \mathbf{t}\cdot\nabla n. \tag{9.25}$$

Equation (9.23) then gives

$$\mathbf{K} = \frac{\nabla n - \mathbf{t}(\mathbf{t}\cdot\nabla n)}{n}. \tag{9.26}$$

The numerical computation then proceeds as follows. Assuming $\mathbf{r}(s_0)$ and $\mathbf{t}(s_0) = \mathbf{r}'(s_0)$ known at the starting point P_0, calculate $\mathbf{K}$ at P_0 from Eq. (9.26). Select an arbitrary small value for Δs and compute $\mathbf{t}(s)$ from Eq. (9.22) using only two terms on the right-hand side. Also calculate $\mathbf{r}(s)$ from Eq. (9.19) using only two terms. By this procedure approximate values of $\mathbf{r}$ and $\mathbf{t}$ are found for a point P a small distance away from the starting point P_0. Iterating this routine determines (approximately) a succession of points of the ray and also the direction of the ray at each point.

This ray-tracing scheme can be made very accurate, at the expense of a longer running time, by taking Δs very small at each step. Also the method is valid regardless of what type of symmetry, if any, is present. However, in component form, three equations for $\mathbf{r}$ and three for $\mathbf{t}$ are involved. Also the calculation of ∇n means that numerical differentiation is required in most cases. Consequently, it is not certain that a single computer program based on this approach is preferable to several separate programs that take account of the special symmetries.

Chapter 10

LENS DESIGN WITH GRADIENTS

10.1 General Comments

Up to the present time lens design with gradients (outside of the fiber optics field) has not received very much attention for two reasons. It was felt that both the fabrication and measurement technology still involved practical difficulties. Furthermore, technical advances in the making of aspheric surfaces have indicated that many design objectives can be accomplished by this means.

On the other hand, the design work that has been done with gradients gives evidence that this approach also has attractive possibilities. It appears that many results that are possible with aspherics can at least be duplicated by means of gradients. Sands (1970), for example, has shown that the contributions of an axial gradient to the third-order aberrations of a system are equivalent to those of an aspheric surface. As evidence of this fact, Moore (1977b) has shown that an axial gradient can be used to replace the aspheric surface in a Schmidt system.

It was already seen that a Wood lens can be designed so as to be completely free of both third-order spherical aberration and astigmatism, something that cannot be done with a conventional homogeneous singlet even when curved surfaces are permitted. If one starts with a Wood lens and then introduces a bending of the surfaces, it is clear that a good-quality singlet with a radial gradient can be designed.

In a system containing inhomogeneous elements, it is found that the paraxial features of the system can be obtained in much the same way as if the elements were all homogeneous. Kapron (1970) considers the case of a convergent Wood lens with a finite object distance, shows how to locate the cardinal points, and describes the general behavior of the paraxial rays in this type of lens.

Sands (1971b) considers first the use of a general singlet with a rotation-symmetric gradient and then specializes to the cases of radial gradient and axial gradient singlets. It is clear that, once the thickness, curvatures, and the gradient parameters have been specified, the paraxial tracing formulas lead quite easily to the paraxial properties of the singlet. It is only necessary to take note of the paraxial form of the refraction at each surface, the effective indices being the axial values N_0 and N'_0 before and after the surface in each case.

By these methods it can be verified, in any rotation-symmetric system containing gradient elements, that most of the familiar paraxial properties of ordinary systems are retained even though the rays in the gradient elements follow curved paths. For example, when any object plane is selected, there exists a unique conjugate image plane such that every point of the first plane is imaged sharply (in the paraxial approximation) at a corresponding point of the second plane. Also, the paraxial magnification is the same for all points of the given object plane.

10.2 A Simple Axial-Gradient Lens

The Wood lens discussed in previous chapters represents the simplest form of radial-gradient singlet. The two plane surfaces provide no

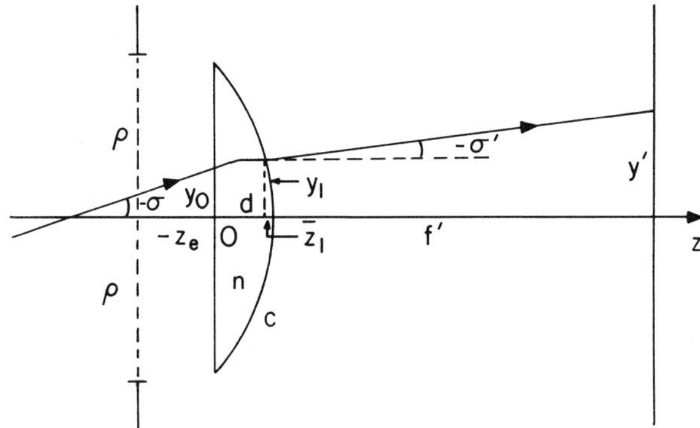

Fig. 10.1. Axial-gradient singlet.

power, that being generated by the gradient. The simplest form of axial gradient singlet must have at least one curved surface if the lens is to have any power. It is therefore instructive to consider an axial-gradient singlet in air with one plane and one spherical surface, as pictured in Fig. 10.1.

In order to make the example as simple as possible, we select an index function that permits integration of the ray-tracing formulas, Eqs. (4.4), simply and exactly. Assuming an index function in the form

$$n = N_0(1 + \alpha z)^{1/2}, \tag{10.1}$$

so that the dielectric constant n^2 is a linear function of z, we find that Eqs. (4.4) reduce to

$$x_1 = x_0 + \frac{2p_0(l_1 - l_0)}{N_0^2 \alpha},$$

$$y_1 = y_0 + \frac{2q_0(l_1 - l_0)}{N_0^2 \alpha}. \tag{10.2}$$

Here the subscripts 0 and 1 indicate values at two points P_0 and P_1 respectively on the ray. As usual for an axial gradient,

$$p_1 = p_0, \qquad q_1 = q_0, \qquad l_1 = (n_1^2 - p_0^2 - q_0^2)^{1/2}. \tag{10.3}$$

As a further simplification, the object plane is assumed to be at infinity, so that the image plane coincides with the focal plane. The refractive index at the vertex of the curved surface is seen to be $N_0 a$, where

$$a = (1 + \alpha d)^{1/2}, \tag{10.4}$$

d being the thickness of the lens.

In the case of paraxial rays

$$l_0 = n_0 = N_0, \qquad l_1 = n_1 = N_0 a, \tag{10.5}$$

so that Eqs. (10.2) reduce to

$$x_1 = x_0 + \frac{2p_0(a - 1)}{N_0 \alpha},$$

$$y_1 = y_0 + \frac{2q_0(a - 1)}{N_0 \alpha} \tag{10.6}$$

or

$$x_1 = x_0 + \bar{p}_0 \bar{d},$$
$$y_1 = y_0 + \bar{q}_0 \bar{d}, \tag{10.7}$$

where

$$\bar{p}_0 = \frac{p_0}{N_0}, \qquad \bar{q}_0 = \frac{q_0}{N_0}, \tag{10.8}$$

and

$$\bar{d} = \frac{2d}{a + 1}. \tag{10.9}$$

The above formulas determine the transfer of rays within the gradient up to the second surface. Snell's law must then be applied to find the directions of the rays (paraxial or otherwise) in the image space. The refraction at the first surface is particularly simple since the optical direction cosines p and q are invariant on refraction at a plane surface.

The simplicity of the present example makes it feasible to derive algebraic formulas for the third-order aberrations of the lens. It is then easy to optimize the lens in any one of several ways. The third-order aberration formulas for this example are listed below, the derivations being outlined in Appendix D.

Spherical Aberration

$$S = \frac{1}{2} f N_1 c^2 h^2 (f\hat{a} - N_1). \tag{10.10}$$

Meridional Coma

$$C = \frac{-3}{2} N_1 c p^2 q_0 \left[1 + \frac{\bar{z}_e}{f} + c\bar{z}_e(f\hat{a} - 1) \right],$$
$$\bar{z}_e = z_e - \frac{\bar{d}}{N_0}. \tag{10.11}$$

Petzval Curvature

$$P = -\frac{f q_0^2}{N_1}. \tag{10.12}$$

Astigmatism

$$A = \frac{1}{2} f q_0^2 [(1 - N_1 c \bar{z}_e)^2 - N_1 f \hat{\alpha} c^2 \bar{z}_e^2].$$ (10.13)

Distortion

$$D = 50 q_0^2 \left\{ \left(1 + \frac{\bar{z}_e}{f}\right)^3 - \frac{(1 + \bar{z}_e/f)\bar{z}_e^2 c}{f} \right.$$
$$\left. + \frac{N_1 \bar{z}_e}{f} \left[\left(c\bar{z}_e - \frac{1}{N_1}\right)^2 - f\hat{\alpha} c^2 \bar{z}_e^2 \right] \right\}.$$ (10.14)

The symbols and abbreviations used here are listed as follows.

$N_0 =$ index just inside the plane surface

$\alpha =$ coefficient in Eq. (10.1)

$d =$ lens thickness

$a = (1 + \alpha d)^{1/2}$

$N_1 = N_0 a =$ index at second vertex

$$\hat{\alpha} = \frac{\alpha}{2a^2}$$

$q_0 = -\sin \sigma$ (for a meridional ray)

$c =$ curvature of second surface

$f =$ focal length

$f' =$ back-focus

$z_e = z$-coordinate of entrance pupil

$$\bar{z}_e = z_e - \frac{\bar{d}}{N_0}$$

$\rho =$ radius of entrance pupil

$h =$ object-ray height in Eq. (10.10).

The focal length and back-focus are equal and given by

$$\frac{1}{f} = \frac{1}{f'} = c(1 - N_1),$$ (10.15)

and the paraxial ray height in the focal plane is

$$y_0' = q_0 f.$$ (10.16)

In order to optimize the lens we note that the design parameters are N_0, α, d, c, and z_e, that is, the two gradient parameters, the lens thickness, the one curvature, and the pupil position. If a desired focal length, or back-focus, has been selected, it is merely necessary that the parameters obey Eq. (10.15).

Spherical aberration can evidently be corrected by choosing α so that

$$f\hat{\alpha} = N_1, \qquad \alpha = \frac{2N_0 a^3}{f},$$ (10.17)

as seen from Eq. (10.10). If this correction is made, Eq. (10.13) for astigmatism reduces to

$$A = \frac{1}{2}fq_0^2(1 - 2N_1 c\bar{z}_e).$$ (10.18)

In this case the astigmatism can be corrected by choosing z_e so that

$$2N_1 c\bar{z}_e = 1.$$ (10.19)

When both of the above corrections have been made, Eq. (10.11) for coma takes the simple form

$$C = \frac{-3}{2} N_1 c\rho^2 q_0,$$ (10.20)

and the distortion formula, Eq. (10.19), reduces to

$$D = \frac{1}{2}(5q_0)^2\left(1 + \frac{1}{N_1} + \frac{2}{N_1^2}\right).$$ (10.21)

Inspection of the formulas for coma and Petzval curvature, Eq. (10.12), shows that these aberrations cannot be removed if the previous correction of spherical aberration and astigmatism has been made. It might appear, from Eq. (10.21), that N_1, the refractive index at the vertex of the curved surface, can be selected so as to remove the distortion. But setting $D = 0$ leads to a value of N_1 that cannot be fabricated.

The following numerical example illustrates the feasibility of a singlet of this type.

Lens Data

aperture	f/5.3
object plane	at infinity
focal length	10 cm
back-focus	10 cm
gradient	$n = N_0(1 + \alpha z)^{1/2}$
gradient parameter	$\alpha = 0.312160711$
first index	$N_0 = 1.5$
index at vertex	$N_1 = 1.52$
thickness	$d = 0.086$ cm
pupil position	$z_e = -1.65$
pupil radius	$\rho = 0.942$ cm
lens diameter	$H = 1.884$ cm
surface curvature	$c = -0.1923077$ cm^{-1}

The half-field for this lens is nearly 50°. However, the extreme rays are subject to considerable vignetting, and therefore we present the third-order aberrations for a nominal field angle of 30°. These are

spherical aberration	$S = 0$
astigmatism	$A = 0$
coma	$C = 0.195$
distortion	$D = 7.9\%$
Petzval curvature	$P = -1.64$

This simple example illustrates the design possibilities with an axial gradient. Although the spherical aberration and astigmatism are completely corrected, the distortion is very large. Of course the restriction of the first surface to a plane reduces the possibility of correcting the aberrations further.

Actually, an axial-gradient singlet of the above type can be made of glass by applying the ion-diffusion technique at the plane surface of a homogeneous singlet of the same shape. The maximum index change required in the present example is given by $N_1 - N_0 = 0.02$ over an interval of $d = 0.086$ cm. This can certainly be achieved by present technology.

It is conceivable that a singlet of the above type could be made of plastic by pouring melted plastic slowly into a spherical mold while gradually changing the index of the entering material.

10.3   Paraxial Rays in a Gradient Singlet

The tracing of paraxial rays in spherical and axial gradients is naturally somewhat simpler than the tracing of real rays. In Eqs. (3.57) and (3.58), however, a quadrature is still required, and the same is true, in general, for an axial gradient as seen from Eqs. (4.4) with l replaced by n. For radial gradients, however, the trace of paraxial rays is particularly easy when the gradient is expressed in the form of Eq. (6.1). It is found from Eqs. (6.4) to (6.8) that the paths of the paraxial rays depend only on the constants N_0 and N_1, apart from the initial conditions. This being the case, the trace through a gradient singlet can be performed without difficulty.

For refraction at the first surface of a singlet in air, the axial values of the refractive indices before and after the surface are unity and N_0 respectively. Application of Snell's law for a paraxial meridional ray leads to

$$q_0 = -y_0\left[\frac{1}{z_0} + c_1(N_0 - 1)\right], \tag{10.22}$$

where $q_0 = -N_0 \sin \sigma_0$, y_0 is the ray height at the surface, c_1 is the surface curvature, and z_0 is the distance of the axial point of the object

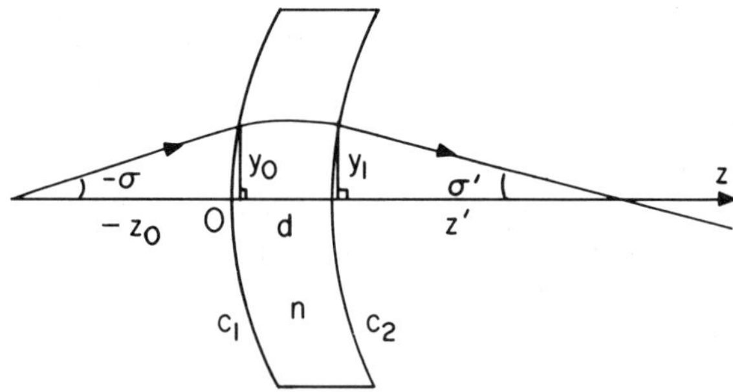

Fig. 10.2.   Radial-gradient singlet.

ray from the first vertex. We define $z_0 \gtrless 0$ as this point is to the right or left of the surface. Thus the direction σ_0 of the ray just after the first surface is determined (see Fig. 10.2).

If $N_1 < 0$, Eqs. (6.5) and (6.6) now give the ray height and direction for the ray as it intersects the second surface, i.e.,

$$y_1 = cy_0 + sq_0/b,$$
$$q_1 = cq_0 - sby_0 = -N_0 \sin \sigma_1,$$

(10.23)

where, from Eq. (6.4),

$$c = \cos \bar{k}, \qquad s = \sin \bar{k}, \qquad \bar{k} = kd = \frac{bd}{N_0},$$

(10.24)

d being the lens thickness.

If $N_1 > 0$, these equations are replaced by

$$y_1 = cy_0 + \frac{sq_0}{b},$$
$$q_1 = cq_0 + sby_0,$$

(10.25)

with

$$c = \cosh \bar{k}, \qquad s = \sinh \bar{k},$$

(10.26)

as seen from Eqs. (6.7) and (6.8).

Paraxial refraction at the second surface gives

$$q' = q_1 + y_1 c_2(N_0 - 1) = -\sigma' = -\frac{y_1}{z'},$$

$$\frac{1}{z'} = -\left[\frac{q_1}{y_1} + c_2(N_0 - 1)\right].$$

(10.27)

These equations, together with the preceding formulas for y_1 and q_1, determine the slope of the image ray and the distance to its axial intersection point.

It is now a simple matter to find the focal length and back-focus in terms of the construction data of the singlet. We consider a ray with

$\sigma = 0$ (i.e., $z_0 = \infty$) and $y_0 = h$ and apply the definition

$$f = \lim_{h \to 0} \frac{h}{\sigma'}. \tag{10.28}$$

When $N_1 < 0$, this leads to

$$\frac{1}{f} = (c_1 - c_2)(N_0 - 1)c + bs$$

$$+ \frac{c_1 c_2 d}{N_0} \frac{(N_0 - 1)^2 s}{\overline{k}}, \tag{10.29}$$

with c, s, and $\overline{k}$ defined by Eqs. (10.24). When $N_1 > 0$, the corresponding formula is

$$\frac{1}{f} = (c_1 - c_2)(N_0 - 1)c - bs$$

$$+ \frac{c_1 c_2 d}{N_0} \frac{(N_0 - 1)^2 s}{\overline{k}}, \tag{10.30}$$

but with c and s here defined by Eqs. (10.26).

If $c_1 = c_2 = 0$ (the Wood lens), these formulas reduce to

$$\frac{1}{f} = \pm bs, \tag{10.31}$$

depending on whether $N_1 \lessgtr 0$, in agreement with Eqs. (8.7) and (8.8). On the other hand, when $b \to 0$, the above focal length formulas reduce to the usual one for a thick homogeneous singlet.

When the object distance z_0 is given by a finite value, it is important to locate the corresponding paraxial image plane. From Eqs. (10.22), (10.23), and (10.27) it follows that, for $N_1 < 0$,

$$\frac{1}{z'} = \frac{c[1/z_0 + c_1(N_0 - 1)] + bs}{c - (s/b)[1/z_0 + c_1(N_0 - 1)]} - c_2(N_0 - 1),$$

$$c = \cos \overline{k}, \qquad s = \sin \overline{k}. \tag{10.32}$$

Likewise, using Eqs. (10.4) instead of (10.2), we have, for $N_1 > 0$,

$$\frac{1}{z'} = \frac{c[1/z_0 + c_1(N_0 - 1)] - bs}{c - (s/b)[1/z_0 + c_1(N_0 - 1)]} - c_2(N_0 - 1),$$

$$c = \cosh \overline{k}, \qquad s = \sinh \overline{k}. \tag{10.33}$$

The back-focus f' in the above cases can be found at once as the value assumed by z' when $z_0 \to \infty$. The result in both cases can be written as

$$f' = f\left[c - \frac{(N_0 - 1)c_1 s}{b}\right],\tag{10.34}$$

where Eqs. (10.29) and (10.30) were used. But, of course, c and s are defined differently in the two cases. Again, in the case of a Wood lens all of the above formulas become very simple.

The corresponding paraxial formulas for an axial-gradient singlet in air are obtained in much the same way but involve a quadrature, as already mentioned. The refraction at the first surface is again given by Eq. (10.22), where $N_0 = n(0)$, the refractive index value at the first vertex.

For the transfer within the gradient we note that paraxially we have $l = n$, so that Eqs. (4.4) give

$$y_1 = y_0 + q_0\hat{d}, \qquad q_1 = q_0,\tag{10.35}$$

$$\hat{d} = \int_0^d dz/n.\tag{10.36}$$

The refraction at the second surface is expressed again by Eq. (10.27) except that N_0 must now be replaced by $N_1 = n(d)$; thus

$$q' = q_1 + y_1 c_2(N_1 - 1),$$

$$\frac{1}{z'} = -\frac{q'}{y_1}.\tag{10.37}$$

Proceeding in the same manner as before we find the focal length to be given by

$$\frac{1}{f} = c_1(N_0 - 1) - c_2(N_1 - 1)$$

$$+ c_1 c_2\hat{d}(N_0 - 1)(N_1 - 1),\tag{10.38}$$

and

$$\frac{1}{z'} = \frac{1/f + (1/z_0)[1 + \hat{d}c_2(N_1 - 1)]}{1 - \hat{d}[1/z_0 + c_1(N_0 - 1)]}.\tag{10.39}$$

It is not difficult to verify that these formulas reduce to the classical ones for a homogeneous singlet if we set $N_0 = N_1 = N$, $\hat{d} = d/N$. Also, Eq. (10.39) gives the correct focal length for the special axial singlet previously considered, in which $c_1 = 0$, $c_2 = c$, and $N_1 = N_0 a$.

The paraxial formulas for a spherical gradient singlet can be obtained with the help of Eqs. (3.57) and (3.58).

10.4 Designing Gradient Singlets

In the design of an inhomogeneous singlet the system parameters include the object distance, the lens thickness, the two lens curvatures, the coefficients needed to describe the gradient (typically three in number), and the location of the aperture stop. If the two surfaces are aspheric, additional parameters are involved. Thus, the design problem is by no means trivial especially since simple formulas for the five third-order aberrations are not, in general, available. Optimizing the lens for third-order aberrations presents a challenge, and higher-order aberrations are of course much more difficult to deal with.

In designing singlets Moore (1971) applied a technique called "double graphing," which takes advantage of the power of the computer to generate many values of the aberrations corresponding to various choices of the system parameters. The double graphing method consists of considering two aberrations at a time with attention to their dependence on the system parameters. In the graph one aberration is chosen as the ordinate and the other as the abscissa.

By the above method Moore was able to design a singlet with an axial gradient that is corrected for third-order spherical and comatic aberrations and has no third-order distortion. In another example, using a radial gradient, he succeeded in correcting all third-order monochromatic aberrations except Petzval curvature. For other examples of gradient singlets optimized for third-order aberrations see Moore (1971) and Moore and Sands (1973).

In the examples described above the singlets were corrected for third-order aberrations. For this purpose only two paraxial rays have to be traced through the system, after which the third-order aberrations can be computed by summing the surface-by-surface and medium-by-medium contributions in accordance with the Buchdahl–Sands theory. This method has the advantage not only of simple ray-tracing requirements but also of supplying information about which system parameters affect the aberrations and by how much.

If, however, higher-order or total aberrations are to be evaluated, it becomes necessary either to trace nonparaxial rays or else to apply very complicated aberration formulas. By using a computer subroutine for solving the differential equations of the rays, a photographic singlet was designed by Marchand and Janeczko (1974) based on optimizing the total aberrations. The resulting lens is considerably better than the best possible homogeneous singlet of the same general type.

This lens, pictured in Fig. 10.2, is of the front-meniscus type with stop located in the image space and the object plane at infinity. The lens, in an f/6.3 version, is corrected for spherical aberration, coma, and tangential and sagittal curvature with a half-field of 25°. The aberrations referred to are total, as defined in Chapter 7, rather than third order, fifth order, etc.

It is interesting to note that the gradient in this lens is of the special type given by Eq. (5.34) with the negative sign present. In Fig. 10.3 curves are shown indicating the spherical aberration and astigmatism, the latter being taken simply as $Z_m - Z_s$, as defined by some authors. Table 10.1 gives lens data and numerical values of the spherical aberration, and Table 10.2 gives ray data and curvature, distortion, and coma values. This lens, operating at f/6.3, roughly matches the performance of the corresponding homogeneous singlet operating at f/11.

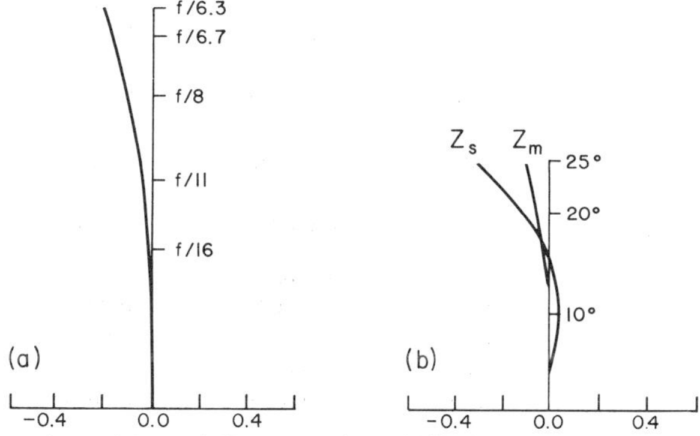

Fig. 10.3. (a) Spherical aberration and (b) astigmatism of gradient singlet (from Marchand and Janeczko, 1974).

TABLE 10.1

Lens Data and Spherical Aberration[a]

Diaphragm openings		Spherical aberration
f/16	1.0	-0.043
f/11	1.2	-0.081
f/8	1.87	-0.153
f/6.7	2.25	-0.222
f/6.3	2.4	-0.253

Refractive index:	Aperture	f/6.3
Radial gradient	Back-focus	$f' = 9.440$
$n^2 = N_0^2 - b^2 r^2$	Focal length	$f = 10.0$
$N_0 = 1.5215$	Half-field	$25°$
$b = 0.1046$	Curvatures	$c_1 = 0.3600$
$\Delta n = -0.008$		$c_2 = 0.1846$
(Max. change in index)	Thickness	$d = 0.45$
	Entrance pupil	$z_e = 3.70$

[a] From Marchand and Janeczko (1974), with a corrected value of c_2.

TABLE 10.2

Meridional Rays, Field Curvature, Distortion, and Coma[a]

	$\Delta y' = y'_p - y'$		
$y_e{}^b$	$\sigma = -25.5°$	$\sigma = -18.0°$	$\sigma = -10.0°$
0	$y'_p = 5.080$	$y'_p = 3.347$	$y'_p = 1.779$
-0.80	$\cdots$	0.033	0.023
-0.75	0.029	0.016	0.011
-0.454	0.014	0.0071	0.0041
-0.333	0.0077	0.0036	0.0017
0.333	-0.0050	-0.0037	-0.0024
0.454	-0.0084	-0.0069	-0.0051
0.625	-0.017	-0.015	-0.012
0.75	-0.025	-0.024	-0.020
0.80	-0.030	-0.028	-0.024
Z_m	-0.093	-0.056	0.002
Z_s	-0.352	-0.103	0.0099
D	6.51%	3.01%	0.89%
C at f/11	0.0027	0.00009	-0.0005
C at f/6.3	0.010	0.0025	-0.00045

[a] From Marchand and Janeczko (1974).

[b] y_e = height of ray at entrance pupil plane.

In a recent paper Moore (1977a) reexamines the design of singlets using either radial or axial gradients. He considers cases with the object plane at infinity similar to the f/6.3 singlet described above and rightly points out that the chief application of such singlets is probably as collimators. This is because it does not appear feasible ordinarily to correct the chromatic aberration in a single-element lens, and therefore good performance can be expected only with monochromatic light such as that from a laser.

In one example, Moore describes an f/4 axial gradient singlet completely corrected for third-order spherical aberration, coma, and distortion. By ray tracing it was found that the fifth-order spherical aberration was also very low. The details of this lens are indicated in Table 10.3. A linear axial gradient was sufficient to achieve a good state of correction. Analysis shows that the tolerance for the index profile is not critical, especially for z values greater than the maximum value of the sag at the first surface.

TABLE 10.3

Axial-Gradient Collimator[a]

Lens data		3rd-order aberration coefficients	
Aperture	f/4	Spherical aberration	0
Focal length	$f = 20.0$ cm	Coma	0
Diameter	$2a = 5.0$ cm	Astigmatism	-0.000488
Field angle	$\tan \sigma = 0.02$	Petzval curvature	-0.000304
Object	at infinity	Distortion	0
First curvature	$c_1 = 0.078496$ cm^{-1}		
Thickness	$d = 1.0$ cm		
Second curvature	$c_2 = 0.001673$ cm^{-1}		
Gradient	$n = 1.65 - 0.0302947z$		

[a] From Moore (1977a).

In a second example, Moore designed a radial gradient singlet with excellent third-order correction. But in this case the fifth-order spherical aberration proved to be excessively large, so that the lens would not appear to be of much practical use.

A third example consists of another radial-gradient singlet with a "shallow" gradient ($N_1 = 0$). By introducing appropriate values of N_2 and N_3 a good correction of the higher-order spherical aberration was achieved. A description of this lens is given in Table 10.4.

TABLE 10.4

Shallow Radial-Gradient Singlet[a]

Lens data		3rd-order aberration coefficients	
Aperture	f/5	Spherical aberration	0
Focal length	$f = 10$ cm	Coma	0
Diameter	$2a = 2$ cm	Astigmatism	-0.04269
Field angle	$\tan \sigma = 0.30$	Petzval curvature	-0.02913
Object	at infinity	Distortion	-0.00646
First curvature	$c_1 = 0.172063$ cm^{-1}		
Thickness	$d = 1.0$ cm		
Second curvature	$c_2 = -0.010393$ cm^{-1}		
Gradient	$n = 1.55 + 0.000261r^4$		
	$\quad + 0.000030r^6$		

[a] From Moore (1977a).

Up to the present time little attention has been paid to the use of spherical gradients in lens design except, of course, for axial gradients, which are obtained as a limiting case. As already pointed out, the ray-tracing equations are somewhat more complicated for spherical gradients, but there is no reason why they should not prove useful in some cases. Furthermore, a spherical gradient can be created rather simply by treating a spherical glass surface by the ion-diffusion method.

10.5   A Spherical-Gradient Singlet

As already pointed out in Chapter 3, the quadrature involved in tracing rays in a spherical gradient can be performed exactly if n^2 is a quadratic function of the distance r from the center of symmetry. Accordingly, it may be helpful to assume this type of gradient profile. The exact algebraic ray-tracing formulas can be examined with a view to choosing the various parameters so as to meet practical constraints and reduce the aberrations. Later, the design can be refined by modifying the gradient profile and using numerical methods and a computer program.

To illustrate this procedure we consider as an example a plano-convex singlet with a spherical gradient near the curved surface as shown in Fig. 10.4. To make the example simple we assume the left-hand part of the lens to be homogeneous of index n_0, the two curved surfaces

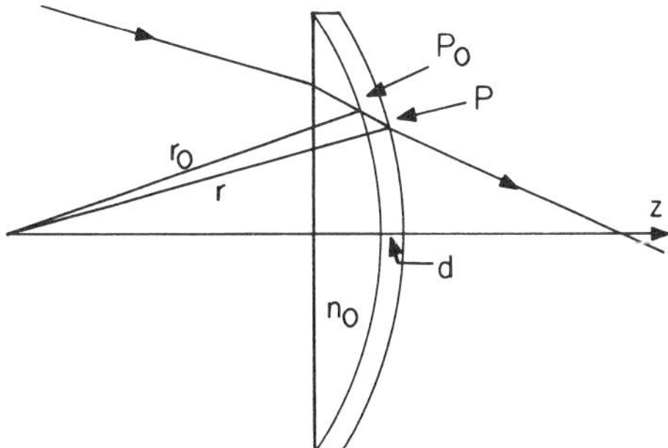

Fig. 10.4.   Spherical-gradient singlet.

to be spheres of constant index n_0 and n, and these to have radii of curvature of $-r_0$ and $-r$ respectively. Thus, there is no index change at the internal interface. We assume the index profile of the gradient to be given by Eq. (3.76), i.e.,

$$n = N_0\left[2 - \delta\left(\frac{r}{r_0}\right)^2\right]^{1/2}. \tag{10.40}$$

The paraxial tracing formulas for rays within the gradient are given by Eqs. (3.57) and (3.58). Here these reduce to

$$M = \int_0^d \frac{du}{n(1 - \rho_0 u)^2}, \qquad d = r - r_0,$$

$$y = \bar{r}(\alpha y_0 + M q_0), \qquad \bar{r} = 1 - \rho_0 d = \frac{r}{r_0},$$

$$q = \frac{1}{r}\left[(\alpha n \bar{r} - n_0) y_0 + (r_0 + M n \bar{r}) q_0\right], \tag{10.41}$$

$$q_0 = -n_0 \sin \sigma_0,$$

$$\alpha = 1 - \frac{n_0 M}{r_0}, \qquad \varepsilon = +1,$$

for a ray in the plane $x = 0$.

Refraction at the second surface is given by

$$q' = q - \frac{(n - 1)y}{r}. \tag{10.42}$$

It is now a simple matter to find the focal length f of the singlet. Taking $q_0 = 0$ and $y_0 = h$, we have, from Eqs. (10.41) and (10.42),

$$\frac{1}{f} = -\lim_{h \to 0} \frac{q'}{h} = -\left[\frac{\alpha n}{r_0} - \frac{n_0}{r} - \frac{(n-1)\alpha}{r_0}\right] \tag{10.43}$$

or

$$\frac{1}{f} = \frac{n_0}{r} - \frac{\alpha}{r_0}$$

$$= \frac{n_0}{r} - \frac{(1 - Mn/r_0)}{r_0}. \tag{10.44}$$

The evaluation of M is quite simple. We first introduce $\bar{r}$ as variable of integration, i.e.,

$$M = r_0 \int_1^{\bar{r}} \frac{d\bar{r}}{n\bar{r}^2}$$

$$= \frac{r_0}{N_0} \int_1^{\bar{r}} \frac{d\bar{r}}{\bar{r}^2 (2 - \delta\bar{r}^2)^{1/2}}$$

$$= \frac{r_0}{N_0} \int_1^{\bar{r}} \frac{d\bar{r}}{\bar{r}^3 (2/\bar{r}^2 - \delta)^{1/2}} \tag{10.45}$$

Then setting

$$v = \frac{1}{\bar{r}^2}, \qquad dv = -\frac{2}{\bar{r}^3} d\bar{r}, \tag{10.46}$$

we have

$$M = -\frac{r_0}{2N_0} \int_1^{v} \frac{dv}{(2v - \delta)^{1/2}}$$

$$= \frac{r_0}{2N_0} \left[(2 - \delta)^{1/2} - (2v - \delta)^{1/2}\right]. \tag{10.47}$$

But from Eq. (10.40)

$$n = N_0 \left(2 - \frac{\delta}{v}\right)^{1/2},$$

$$n_0 = N_0 (2 - \delta)^{1/2}, \tag{10.48}$$

so that

$$M = \frac{1}{2}\left(\frac{r_0}{N_0}\right)^2\left(\frac{n_0}{r_0} - \frac{n}{r}\right).$$
(10.49)

Hence, from Eq. (10.44),

$$\frac{1}{f} = \frac{n_0}{r} - \frac{1}{r_0} + \frac{n_0}{2N_0^2}\left(\frac{n_0}{r_0} - \frac{n}{r}\right)$$
(10.50)

for the paraxial focal length.

We can verify that this agrees with the well-known result for a homogeneous plano-convex singlet. When $\delta = 0$,

$$n = n_0 = N_0\sqrt{2},$$
(10.51)

in which case Eq. (10.50) reduces to

$$\frac{1}{f} = \frac{(n_0 - 1)}{r}$$
(10.52)

as it should.

The back-focus f' is easily found by considering the same ray as before and using the relation

$$\frac{h}{f} \simeq \frac{y}{f'}$$
(10.53)

or

$$f' = f \lim_{h \to 0} \frac{y}{h} = f\alpha\bar{r}.$$
(10.54)

We shall not examine this example further. However, it should be noted that the equations for exact ray tracing are available in Chapter 3, including Eq. (3.83) with $\varepsilon = +1$. In the present example, the usual iterative solution of Eq. (3.29) to determine u is completely avoided as we have simply

$$u = d = r - r_0$$
(10.55)

at the point where the ray intersects the outer surface.

10.6   Replacing Aspherics by Gradients

It has been known for some time that the third-order surface con-tribution of an inhomogeneous medium to the aberration polynomial is the same as that of an aspheric surface (see Sands, 1970). It appears, therefore, that an important potential application of gradients lies in the possibility of utilizing them in place of aspheric surfaces in high-performance optical systems. It seems probable that most design objec-tives that can be achieved through the use of aspherics can be matched by using gradients having spherical surfaces. If this proves to be the case, the choice in practice naturally will depend on the relative cost and difficulty of producing optical elements of these two types.

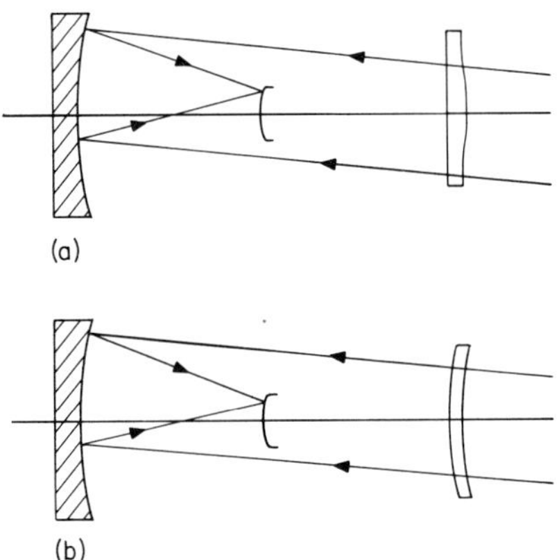

Fig. 10.5. Schmidt system (a) with aspheric corrector plate and (b) with gradient corrector plate.

Perhaps the first attempt to exploit this idea is that of Moore (1977b), who has considered the possibility of replacing the aspheric corrector plate in a Schmidt telescope by an axial gradient element. Figure 10.5 shows the arrangement of a conventional Schmidt system (a) with an aspheric corrector plate and (b) with a gradient index corrector plate. Light from a distant object point is shown entering from the right. After passing through the corrector plate it is reflected off the primary mirror and converges at a point of the focal surface. Due to the curvature of the focal surface an additional lens element is often inserted to flatten the field.

The design of the gradient corrector plate presents no particular difficulties. In fact, the coefficients of the gradient function can be related directly to those in the equation of the corresponding aspheric surface. With only a linear gradient function and one planar and one spherical surface for the corrector plate the third-order spherical aberration of the system can be corrected. However, the question of chromatic aberration is important, the paraxial chromatic aberration being directly proportional to the power of the plate and inversely to the v value of the glass. The power of the corrector plate can be reduced sharply by having the curvature of its two surfaces nearly equal, these curvatures being taken as nonzero to effect the correction of spherical aberration. Thus, a good correction of the chromatic aberration can be achieved.

The correction of higher-order aberrations must also be considered. These can be reduced somewhat by introducing higher-order terms in the gradient function. Moore gives the specifications of a number of possible corrector plates that can be used with an f/2.5 primary mirror of diameter 15 cm. The particular choice depends then on which can be manufactured. It appears, however, that these examples do lie within the capabilities of present technology.

This study of the Schmidt camera indicates that a gradient element can, indeed, be used in place of an aspheric element in this application. The resulting performance is at least equivalent to that achieved in the past with an aspheric surface.

10.7   A Simple Image Inverter

It is often necessary to include in an optical system an element that turns an inverted image right-side up. Such is the case with binoculars, the terrestrial telescope, and certain types of viewfinders in cameras. Methods based on prisms and also "twisters," consisting of bundles of homogeneous fibers arranged in a twisted pattern, have been applied. However, a suitably designed GRIN rod can serve the same purpose.

The GRIN rod is essentially the same as a long Wood lens except that the object surface is usually located at the first surface and the image plane at the second surface. Let the index function be represented by

$$n = N_0 + N_1 r^2 + N_2 r^4, \tag{10.56}$$

where r is the distance from the z-axis. The paraxial tracing formulas then are given by Eqs. (10.23) and (10.24) for a ray in the plane $x = 0$ and terminating at the distance $z = d$, i.e.,

$$x_1 = cx_0 + \frac{sp_0}{b},$$

$$y_1 = cy_0 + \frac{sq_0}{b} \tag{10.57}$$

with

$$c = \cos \bar{k}, \qquad s = \sin \bar{k}, \qquad \bar{k} = kd = d|2N_1/N_0|^{1/2} \tag{10.58}$$

holding if $N_1 < 0$.

If d is chosen so that $\bar{k} = \pi$, we have $c = -1$, $s = 0$, in which case

$$x_1 = -x_0, \qquad y_1 = -y_0. \tag{10.59}$$

Thus, every object point at the first surface is sharply imaged at the second surface but with magnification -1 (within the paraxial approximation). Thus, an image present in the first surface appears inverted on the second. The distance d required for this to hold is given by Eq. (10.58) as

$$d = \frac{\pi}{k}, \tag{10.60}$$

with k determined by the ratio N_1/N_0.

In order to improve the image formation we can consider the behavior of nonparaxial rays. For example, spherical aberration of all orders will be corrected if all rays from the axial point of the first surface meet at the axial point of the second surface. Since these are meridional rays, the analysis in Section 5.4 can be applied.

All meridional rays will have the same period if the refractive index profile is of the form of Eq. (5.48), i.e.,

$$n = N_0 \operatorname{sech}(\alpha r), \tag{10.61}$$

where N_0 and α are constants. Inspection of the solution of the tracing equation for meridional rays, as given by Eq. (5.53), shows that the half-period is equal to π/α. It follows that, if $\alpha = k$, all paraxial rays and also all meridional rays are sharply imaged with magnification -1 at the distance $d = \pi/k$.

It appears, therefore, that Eq. (10.61) with $\alpha = k$ is the optimal choice for the index profile. By series expansion we see that this function can be approximated by

$$n = N_0 \operatorname{sech}(kr)$$

$$= N_0\left[1 - \frac{1}{2}(kr)^2 + \left(\frac{5}{24}\right)(kr)^4\right], \qquad (10.62)$$

showing, by comparison with Eq. (10.56), that we should choose

$$N_2 = \left(\frac{5}{24}\right)N_0 k^4. \qquad (10.63)$$

As a numerical example, consider a rod of diameter 1 cm with $N_0 = 1.5$ and $N_1 = -0.2$. This calls for an index change of approximately 0.05 in a radial distance of 0.5 cm, as seen from the first two terms of Eq. (10.56). This result can be achieved with present technology. In this case $k = 0.5164$, $d = 6.084$ cm, and, from Eq. (10.63), we find that $N_2 = 0.02222$.

It is evident that the above analysis applies for the design of a GRIN rod to be used to produce an erect image. The only difference is that d should be taken as $d = 2\pi/k$, or any integral multiple of that quantity, to produce an erect image.

A study of aberrations of Selfoc fibers was made by Brushon (1975). He considers a Selfoc fiber to be the same as a GRIN rod but with object plane outside the fiber.

Chapter 11

FABRICATION OF GRADIENT ELEMENTS

11.1 The Diffusion Method

Reference has already been made to the experiments of R. W. Wood around 1900, in which he was able to make simple radial-gradient lens elements of gelatin. About the same time the German glassmaker Schott produced weak gradients in glass. His method was to pour molten glass into cylindrical iron molds, after which rapid chilling from the outside inward produced the index variation. The Schott Company still makes gradient elements but almost certainly not by the above method.

Extensive studies have been made of gas lenses and other media to be used for the propagation of coherent light in waveguides (see Marcuse and Miller, 1964; Marcuse, 1965a,b; Miller, 1965; Berreman, 1965; Marcatili, 1967; Kornhauser and Yaghjian, 1967; Streifer and Kurtz, 1967; Kawakami and Nishizawa, 1968). More recently Pearson *et al.* (1969) reported the fabrication of gradient index glass rods for use as image relays. The technique of diffusive ion exchange was applied to produce the gradients.

The technology required for making gradients for image relays is not entirely similar to that for the waveguide application. A waveguide fiber should have extremely low absorption and scattering losses, but the requirement for the refractive index profile is not very restrictive. For the image-relay fiber, rather precise control of the index profile is required, and moderate absorption and scattering losses can be tolerated. Clearly the same is true for gradients to be used in classical lens elements, but this application usually calls for index changes extending over larger distances.

In the same year Hamblen (1969) received a patent for a similar ion-diffusion method of producing gradients in glass and constructing lenses with them. Actually, ion diffusion into glasses has been studied by experimentalists for over fifty years (see, for example, Schulze, 1913, 1915, 1919; Zachariasen, 1932a,b). However, only in the last few years has it been possible to create substantial changes in the refractive index without causing harmful effects such as strain and discoloration.

The diffusion procedure consists of immersing the glass in a molten salt bath at a high temperature, the time period being typically many hours long. During this time cations migrate from the bath and are exchanged with the alkali ions inside the glass.

A cation exchange of this kind is known to be one-for-one, that is, in a binary diffusion each monovalent cation entering the glass is replaced by one alkali–glass ion, which in turn diffuses out of the glass and into the melt.

Factors that influence the rate of exchange are (1) the electrochemical activities between opposing ions, (2) the bonding strengths that retain the alkali–glass ions in their lattice sites, and (3) the relative mobilities of the various ionic species inside the glass network. Ion mobility, in turn, depends on the temperature. A high temperature can be thought of as supplying additional vibrational energy to the glass lattice and, in effect, expanding channels through which the ions are free to move. Although temperature plays a dominant role, the glass composition is important in determining what ions can be exchanged and what the physical condition of the glass will be afterwards.

To increase the index of refraction, alkali–glass ions are exchanged for monovalent cations having a larger ionic radius, i.e., ions having a greater degree of polarization for incident light. Silver and thallium, for instance, are two ions which readily exchange for the smaller ions of sodium and potassium. Alkali–glass ions are loosely bound as proved by studies on electrical conduction, where it is revealed that the mobile alkali ions are responsible for charge transport.

In the diffusion process a nonequilibrium condition exists when a chemical force is set up across the glass interface separating the molten salt from the lattice ions. Complete equilibrium is not allowed, for before this occurs the glass is withdrawn from the bath and cooled, locking the glass ions in place. A concentration gradient of counterions then remains, resulting in a refractive index gradient extending a certain distance from the surface into the glass.

Glass fracture due to introducing large ions into the structure is prevented by conducting the diffusion at a high temperature, allowing the glass, in effect, to be self-annealing. What is more, diffusion into glass is most rapid when performed within the transformation region, that is, the temperature range where the material makes a transition from the glassy state to that of a highly viscous state but below the melting range. At a temperature above 600°C, cation penetration is rapid, with the diffusion coefficients being three to five orders of magnitude greater than their values at normal temperatures.

In carrying out the diffusion procedure it is beneficial to add suitable amounts of sodium and potassium oxides. This lends flexibility in controlling the maximum index change, the slope of the curve, and the depth of penetration of the ions. If it is desired to lower rather than raise the refractive index, this can be accomplished by suitable choice of the glass and the type of cations to be diffused into it.

Some typical results obtained by Hamblen using the diffusion method are shown in Fig. 11.1. Three of the curves show respectively the index profiles obtained by diffusing thallium ions, silver ions, and a combination of these two ions into a sodium–aluminum silicate glass. The fourth curve is for the diffusion of silver ions into the disilicate glass, $Na_2O \cdot 2SiO_2$. These diffusions were conducted at 600°C for 18 hours.

Results similar to those of Hamblen have been reported by Kita and Uchida (1970) in connection with the production of the Selfoc fibers used for image relays. Selfoc is a registered trade name, suggesting a self-focusing device. These workers were able to prepare fibers in which the refractive index is very nearly parabolic as a function of the radial distance. Further developments in the technology of making gradient elements of glass have been reported by Matsushita *et al.* (1972) and by Kitano *et al.* (1972).

The problem of controlling this ion-diffusion process so as to produce gradients having a specified index profile was considered by Martin (1975). As examples, he investigated the three radial gradient functions

$$n = N_0\left(1 - \frac{\alpha^2 r^2}{2}\right), \tag{11.1}$$

$$n = N_0 \operatorname{sech}(\alpha r), \tag{11.2}$$

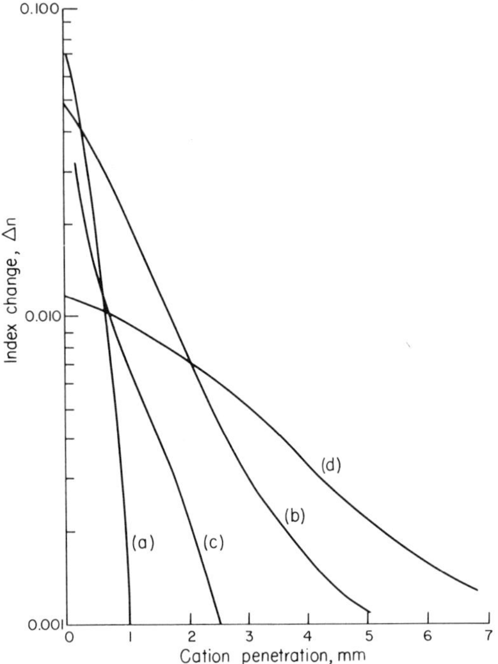

Fig. 11.1. Typical index profiles obtained by ion diffusion: (a) thallium ions into sodium–aluminate glass; (b) same with silver ions; (c) same with thallium and silver ions; and (d) silver into disilicate glass.

and

$$n = N_0(1 + \alpha^2 r^2)^{-1/2}, \tag{11.3}$$

these being cases that have been considered important in the image-relay application of gradient index fibers.

As with Martin's work, most of the research on gradient fabrication has been devoted to the making of fibers either for the waveguide or image-relay applications. Although our chief interest in the present book is not primarily centered on these uses of gradients, it is expected that some of the technology developed for fibers can be modified or adapted to the making of classical gradient optical elements. In principle, for example, a fiber could be cut perpendicular to the axis, making elements that could be made into lenses. However, the small diameter of most fibers presents a serious handicap. In any case we cite several further references describing gradient fiber fabrication technology.

11.2 Other Methods

A method of making optical waveguides by low-energy electron irradiation of silica is described by Houghton and Townsend (1976). The preparation of graded index fibers of silica is also described by O'Connor *et al.* (1976). Techniques for making gradient fiber waveguides with a borosilicate composition are presented by French *et al.* (1976) (see also Cohen, 1976). One objective of theirs was to produce fibers with very low dispersion.

A description of the preparation of graded index fibers in the alkali germanosilicate system is given by Van Ass *et al.* (1976). Scientists at the Corning Glass Works reported a "New glass system for low-loss optical waveguides" (Sommer *et al.*, 1976). Some of the preparations are based on a vapor-deposition technique. A fabrication procedure based on reactive sputtering to make slab waveguides is described by Cheng *et al.* (1975). Planar gradient optical waveguides made by ion implantation in zinc telluride are discussed by Valette *et al.* (1977).

Most of the above references deal with the formation of gradients in glass. One further technique for doing this is of considerable theoretical interest, namely, the forming of gradients in glass by neutron irradiation. Sinai (1971) was able to produce radial-gradient glass singlets of large diameter by irradiation with high-energy neutrons. A wide collimated beam of neutrons is directed at the lens in a direction parallel to the axis. Suitable cadmium masks, however, are placed in front of the lens so as to modify the intensity profile of the incident beam. Thus, in principle, almost any form of radial index gradient can be generated in the lens element.

Considerable progress has been made in the fabrication of gradient index elements of plastic. R. Moore (1971, 1973) has made samples of plastic Wood lenses, and these were found, in general, to have the somewhat surprisingly good imaging qualities that are predicted by theory.

Ohtsuka (1973) and Iga and Yamamoto (1977) have described the fabrication of plastic fibers for imaging applications (see also Iga *et al.*, 1975).

One basic approach used to produce the gradient fibers is to place a polymerized soft plastic rod with a higher refractive index in a bath of a lower-index monomer. Under the influence of elevated temperature a monomer exchange diffusion process results in a nearly parabolic index distribution. Rod samples of 1- to 3-mm diameter and 150-mm

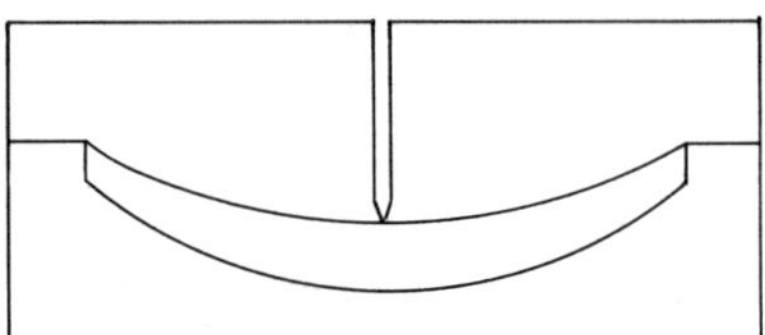

Fig. 11.2.   Setup for making gradient element by diffusion of plastics.

long, for example, can be made, and these are suitable for various applications such as for medical equipment. With the above dimensions a resolution of 5 lines per millimeter has been achieved in the imaging. To accomplish this, careful control of the r^4 term in the radial gradient function is required.

Hamblen (1977) has developed a method of making plastic gradient index lens elements having plane or curved surfaces and relatively large diameters. He uses a rotating silicone rubber mold into which the first monomer enters through a lid via hypodermic injection. Half the mold volume is filled while this liquid is retained in the outside of the lens cavity by centrifugal force (see Fig. 11.2). A second monomer is then injected into the remaining central volume, and the two liquids, one of high and the other of low refractive index, diffuse together to establish a radial gradient. For a converging lens the high index is centralized along the optical axis. Control of the gradient profile is achieved by using ultraviolet light on the transparent mold through a variable density mask, causing partial gelation of the monomers in the regions of maximum light flux.

Chapter 12
MEASUREMENT OF INDEX GRADIENTS

12.1 An Approximate Method

Closely related to the problem of making gradient elements, particularly with careful control of the gradient function, is that of accurate measurement of the gradient in actual samples. A number of methods for doing this have been proposed, but the problem is still a challenging one. The choice of method depends partly on the geometrical configuration of the sample, but also on the degree of precision required. As has been mentioned, for example, the form of the index profile is less critical in the waveguide application than it is in the image-relay or direct imaging applications.

In a radial-gradient singlet, a relatively straightforward approach appears to be feasible for determining the coefficient N_1 in Eq. (6.1) if N_0 and d have already been determined along with the two surface curvatures of c_1 and c_2. The idea is to measure the paraxial back-focus f' when a narrow collimated beam along the axis is incident on the lens. Then implicit formulas for computing N_1 must be solved by numerical methods.

From Eq. (10.32) with $z_0 = \infty$, we obtain, for the case $N_1 < 0$,

$$\frac{1}{f'} = \frac{c_1(N_0 - 1) - bt}{1 - c_1(N_0 - 1)t/b} - c_2(N_0 - 1) \tag{12.1}$$

with

$$t = \frac{s}{c} = \tan \bar{k}, \qquad \bar{k} = d|2N_1/N_0|^{1/2}, \qquad b = \frac{N_0\bar{k}}{d}. \tag{12.2}$$

These equations can be solved numerically for $\bar{k}$, after which N_1 is easily found.

The corresponding formulas for the case $N_1 > 0$ are obtained from Eqs. (10.33). This simple method appears attractive, but it is not clear that sufficient experimental accuracy can be obtained to make the method practical except, perhaps, as a rough check on other methods.

In the case of a Wood lens, the formulas involved in the above method are particularly simple, i.e.,

$$\frac{1}{f'} = b \tan \bar{k} = \frac{N_0\bar{k}}{d} \tan \bar{k}, \tag{12.3}$$

if $N_1 < 0$, and

$$\frac{1}{f'} = -\frac{N_0\bar{k}}{d} \tanh \bar{k}, \tag{12.4}$$

if $N_1 > 0$.

12.2 Interferometric Methods

For improved accuracy usually an interferometric measurement method must be applied. As is the case with fabrication, most of the research on gradient measurement techniques has been devoted to the fiber applications for either waveguide or image-relay purposes.

A basic method that can be considered in almost any application is to make a thin section with optically flat parallel faces and with the index varying with distance along the faces. A monochromatic beam is split so that part passes through the sample and part bypasses it. When the two beams are later brought together, interference fringes indicate the light-path difference related to the local refractive index at various positions in the sample.

One simple arrangement for carrying out this scheme, as suggested by Hamblen, is to use light reflected from the front and back surfaces of the sample as shown in Fig. 12.1. Here an incident beam directed perpendicular to the face of the sample passes through a partial mirror as shown. The returning beams from the front and back faces of the sample interfere with each other either constructively or destructively depending on the optical path in the sample.

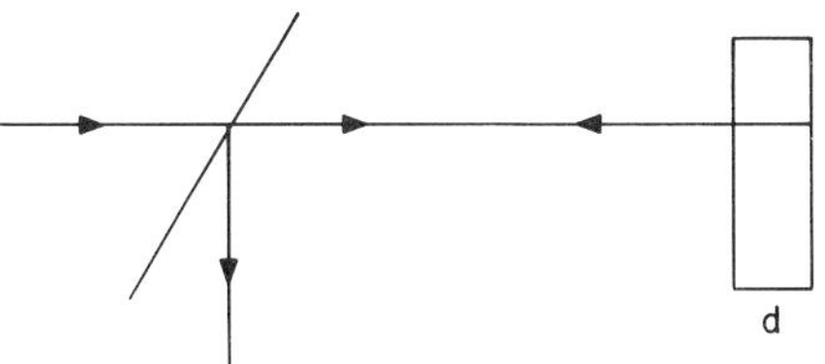

Fig. 12.1.   Gradient measurement arrangement.

The returning beam is reflected from a partial mirror to a receiving plane, which could be a photographic plate. Since the optical path in the sample is proportional to the local refractive index, the spacing of interference fringes at the receiving plane gives an accurate determination of the refractive index gradient in the sample. Each fringe corresponds to an index change of $\lambda/2d$. Some typical curves indicating the dispersion in a plastic gradient sample are shown in Fig. 12.2.

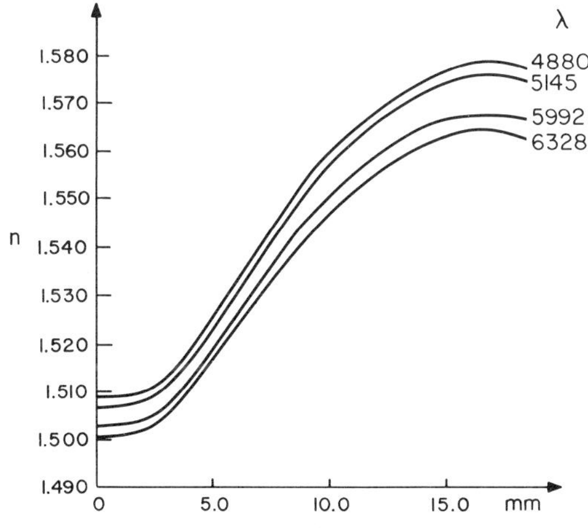

Fig. 12.2.   Typical dispersion curves.

The above measurement method, of course, is not without difficulties, requiring, as it does, the preparation of a thin sample with optically flat parallel surfaces.

A second method for measuring the gradient in a thin sample depends on the use of a modified Abbe interferometer. This approach is based on the measurement of the critical angle for total internal reflection given by the well-known formula

$$\sin \theta_{\mathrm{c}} = \frac{n}{n_1}, \tag{12.5}$$

where n_1 and n are the internal and external refractive indices, respectively.

The essence of this procedure is to send a narrow beam of light through the sample and then rotate the sample until total internal reflection occurs. By measuring the angle of rotation the local refractive index is determined. Provision must be made for translation of the sample so as to measure the index at various points. In a typical instrument of this type, samples up to 1-mm thick can be measured with spatial resolution of 0.1 mm and with index accuracy of 5×10^{-4} over the visible spectrum.

Foelster and Herrmann (1974) developed a method of measuring radial gradients in thin samples by means of an interference microscope. By their method the index function can be determined up to the third term in the series expansion given by Eq. (6.1). For thicker samples, they are able to determine the index only up to the second term. Somewhat related to the above method is that of Martin (1974) (see also Stone and Burrus, 1975) for the measurement of slices of gradient fibers made for the waveguide application. He also uses an interference microscope but applies a reflection technique in the analysis of gradient glass fibers formed by diffusion.

In the case of a gradient fiber the direct interferometric methods appear to have drawbacks. The small radius and small variation in refractive index result in few observable fringes. Also there is some difficulty with the passage of rays twice through a sample. Even a ray parallel to the axis of a radial gradient is curved within the sample and, after reflection at the second surface, returns in a shifted position.

In view of the above difficulties a different method for gradient measurement was introduced by Ikeda *et al.* (1975). This technique consists of measuring the power reflected from various points of the plane end of a fiber and evidently avoids the problem of having to prepare a thin section with optically flat parallel surfaces.

In several of the methods based on analyzing interference fringes the analysis of the results can become tedious and time consuming, especially if large numbers of measurements must be made, as would occur in quantity production of fibers. Thus, attempts have been made to automate the viewing and analysis of the fringe patterns. This can be done using a microdensitometer and also appropriate data reduction (see, for example, Wonsiewicz *et al.*, 1976).

A method of this sort has been devised by a group at the University of Rochester under the direction of D. Moore. The chief interest of this group is in classical gradient lenses, and a systematic routine for making and measuring gradient elements with the ion-diffusion technique has been developed.

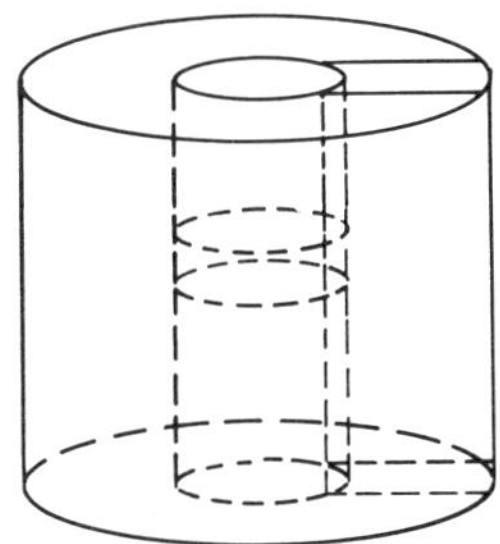

Fig. 12.3.   Making and measuring axial gradients.

An interesting way of producing and measuring axial gradients is used by this group, as indicated in Fig. 12.3. A glass cylinder is immersed in a salt bath and left in a high-temperature oven for many hours. After the gradient has been formed by penetration of the ions into all surfaces of the cylinder, including the ends, the cylinder is cut in the several ways indicated in the figure. In this fashion a central cylinder with an axial gradient is obtained from which lens blanks can be cut. The parallel vertical cuts at the right produce a slab that can be used to measure the axial gradient that has been created. A computer routine fits the measured data to a high-order polynomial in the axial distance.

Recently a number of methods have been developed for measuring the gradient profiles of fibers by analyzing light that is incident on the fiber in a direction perpendicular to its axis of symmetry. Such a procedure is described by Marhic *et al.*, (1975). Also Ohtsuka and Shimizu (1977) describe a technique of this kind based on the shearing method of Interphako microscopy. Interphako is a registered trade name of the firm of Carl Zeiss in Jena, East Germany. These methods involve placing a relatively long section of the fiber in a glass cell filled with an immersion liquid with index matching that at the outer surface of the fiber. By measuring the light path through the fiber at various heights in the beam the gradient index function can be deduced. In some typical examples, these authors have determined values of the coefficients in Eq. (6.1) up to the term in r^8, indicating that a high degree of precision can be achieved for this method.

In connection with the above measurement method the mathematical theory discussed in Chapter 2 can be helpful, for a ray incident on a radial-gradient fiber in a direction perpendicular to the axis obeys the same laws as a ray in a spherical gradient.

A measurement method somewhat similar to that described above has been studied by Saunders and Gardner (1977) for clad fibers with an index profile that can be approximated by the formula

$$n = N_0 \left[1 - 2\Delta \left(\frac{r}{a} \right)^\alpha \right]^{1/2}, \qquad 0 \le r \le a, \tag{12.6}$$

where N_0, Δ, a, and α are constants. This form of function has been referred to as the Gloge–Marcatili profile (see Gloge and Marcatili, 1973) and has been shown to be optimal for the waveguide application.

The measurement method of Saunders and Gardner is designed for speed and simplicity rather than high accuracy and appears to be sufficiently precise for the waveguide application. A simplified variation of their method has been proposed by Stone (1977). Here the refractive index of the immersion fluid is tuned until a specified fringe pattern is matched.

Still other methods for measuring gradient index profiles have been proposed, such as using far-field diffraction patterns (see, for example, Marcuse and Presby, 1975); Maeda and Hamasaki, 1977). A method based on moiré patterns was suggested over thirteen years ago (Nishijima and Oster, 1964). A number of workers determine the index profile of a fiber from the back-scattered light generated when the fiber is illuminated from the side (see Presby, 1974; Watkins, 1974; Okashi and Hotate, 1976; Brinkmeyer, 1977).

Appendix A
DERIVATION OF
EQUATIONS (1.8)

If Eq. (1.2) is written in the form

$$L = \int_{z_0}^{z} n(1 + \dot{x}^2 + \dot{y}^2)^{1/2}\, dz, \qquad (A.1)$$

with a dot indicating d/dz, the Euler equation for x is

$$\frac{d}{dz}\left[\frac{n\dot{x}}{(1 + \dot{x}^2 + \dot{y}^2)^{1/2}}\right] = (1 + \dot{x}^2 + \dot{y}^2)^{1/2}\frac{\partial n}{\partial x}. \qquad (A.2)$$

Carrying out the differentiation and multiplying by $(1 + \dot{x}^2 + \dot{y}^2)^{1/2}$ gives

$$\ddot{x} + \dot{m}\dot{x} - \frac{\dot{x}(\dot{x}\ddot{x} + \dot{y}\ddot{y})}{1 + \dot{x}^2 + \dot{y}^2} = (1 + \dot{x}^2 + \dot{y}^2)\frac{\partial m}{\partial x}, \qquad (A.3)$$

where

$$m = \log n. \qquad (A.4)$$

Rearranging this further we find

$$\dot{x}\dot{y}\ddot{x} - \ddot{y}(1 + \dot{x}^2) = (1 + \dot{x}^2 + \dot{y}^2)\left[m\dot{y} - \frac{\partial m}{\partial y}(1 + \dot{x}^2 + \dot{y}^2)\right]. \qquad (A.5)$$

The analogous equation with x and y reversed is

$$\dot{x}\dot{y}\ddot{y} - \ddot{x}(1 + \dot{y}^2) = (1 + \dot{x}^2 + \dot{y}^2)\left[m\dot{x} - \frac{\partial m}{\partial x}(1 + \dot{x}^2 + \dot{y}^2) \right]. \quad (A.6)$$

Solving these as simultaneous equations for $\ddot{x}$ and $\ddot{y}$ gives

$$\ddot{x} = (1 + \dot{x}^2 + \dot{y}^2)\left[(1 + \dot{x}^2)\frac{\partial m}{\partial x} + \dot{x}\dot{y}\frac{\partial m}{\partial y} - \dot{m}\dot{x} \right] \quad (A.7)$$

with a similar equation for y. However,

$$\dot{m} = \dot{x}\frac{\partial m}{\partial x} + \dot{y}\frac{\partial m}{\partial y} + \frac{\partial m}{\partial z}. \quad (A.8)$$

Inserting this into Eq. (A.7) leads to

$$\ddot{x} = (1 + \dot{x}^2 + \dot{y}^2)\left(\frac{\partial m}{\partial x} - \dot{x}\frac{\partial m}{\partial z} \right), \quad (A.9)$$

which is seen to be equivalent to the first of Eqs. (1.8) if Eq. (A.4) is used.

Appendix B
VERIFICATION OF
TRACING FORMULAS FOR
A SPHERICAL GRADIENT

It is an interesting exercise to show that Eqs. (3.42) and (3.27) satisfy the check formulas (3.45) and (3.46) at every point of a ray. From Eqs. (3.42) and (3.30)

$$
\begin{aligned}
p^2 + q^2 + l^2 &= A^2[\rho_0^2(x_0^2 + y_0^2) + a^2] + B^2[p_0^2 + q_0^2 + l_0^2] \\
&\quad + 2AB[|\rho_0|(x_0 p_0 + y_0 q_0) + a l_0] \\
&= A^2 + n_0^2 B^2 + 2AB n_0 \cos \psi_0.
\end{aligned} \tag{B.1}
$$

Hence

$$
p^2 + q^2 + l^2 = \frac{\overline{A}^2 + \overline{B}^2 + 2\overline{A}\,\overline{B}\cos \psi_0}{(1 - \rho_0 u)^2}, \tag{B.2}
$$

where, from Eqs. (3.43),

$$
\begin{aligned}
\overline{A} &= \alpha t - n_0 \cos(\psi_0 - \theta), \\
\overline{B} &= \beta t + n_0 \cos \theta.
\end{aligned} \tag{B.3}
$$

However, Eqs. (3.44) and (3.25) give

$$t = \frac{\bar{e}}{\tan \psi} = \frac{n_0 \sin \psi_0}{\tan \psi}, \tag{B.4}$$

and so, from Eqs. (3.36),

$$\alpha t = \frac{n_0 \sin(\psi_0 - \theta)}{\tan \psi},$$

$$\beta t = \frac{n_0 \sin \theta}{\tan \psi}. \tag{B.5}$$

It follows that

$$\bar{A} = n_0 \left[\frac{\sin(\psi_0 - \theta)}{\tan \psi} - \cos(\psi_0 - \theta) \right]$$

$$= \frac{n_0 \sin(\psi_0 - \theta - \psi)}{\sin \psi} \tag{B.6}$$

and

$$\bar{B} = n_0 \left[\frac{\sin \theta}{\tan \psi} + \cos \theta \right]$$

$$= \frac{n_0 \sin(\theta + \psi)}{\sin \psi}. \tag{B.7}$$

Accordingly, writing

$$\theta + \psi = \hat{\theta}, \tag{B.8}$$

we have

$$p^2 + q^2 + l^2 = \frac{n_0^2}{(1 - \rho_0 u)^2 \sin^2 \psi} \times \left[\sin^2(\psi_0 - \hat{\theta}) \right.$$

$$\left. + \sin^2 \hat{\theta} + 2 \cos \psi_0 \sin(\psi_0 - \hat{\theta}) \sin \hat{\theta} \right]. \tag{B.9}$$

Equations (3.9), (3.22), and (3.25) show that

$$\bar{e}^2 = n^2(1 - \rho_0 u)^2 \sin^2 \psi = n_0^2 \sin^2 \psi_0, \tag{B.10}$$

from which the first factor in Eq. (B.9) becomes

$$\frac{n_0^2}{(1 - \rho_0 u)^2 \sin^2 \psi} = \frac{n^2}{\sin^2 \psi_0}.$$ (B.11)

Our task, therefore, is to show that, at any point of the ray,

$$\sin^2(\psi_0 - \hat{\theta}) + \sin^2 \hat{\theta} + 2 \cos \psi_0 \sin(\psi_0 - \hat{\theta}) \sin \hat{\theta} = \sin^2 \psi_0.$$ (B.12)

But, by trigonometric manipulations, it is shown that this equation is identically satisfied for all values of ψ_0 and $\hat{\theta}$.

To show that the tracing formulas given by Eqs. (3.27) and (3.42) satisfy Eq. (3.46) it is convenient to use the properties of determinants. Thus

$$\begin{aligned}
\begin{vmatrix} x & p \\ y & q \end{vmatrix} &= (1 - \rho_0 u) \begin{vmatrix} ax_0 + p_0 M \operatorname{sinc} \theta, & A|\rho_0|x_0 + Bp_0 \\ ay_0 + q_0 M \operatorname{sinc} \theta, & A|\rho_0|y_0 + Bq_0 \end{vmatrix} \\
&= (1 - \rho_0 u) \left\{ \alpha B \begin{vmatrix} x_0 & p_0 \\ y_0 & q_0 \end{vmatrix} + |\rho_0| A M \operatorname{sinc} \theta \begin{vmatrix} p_0 & x_0 \\ q_0 & y_0 \end{vmatrix} \right\} \\
&= (1 - \rho_0 u)(\alpha B - A|\rho_0|M \operatorname{sinc} \theta) \begin{vmatrix} x_0 & p_0 \\ y_0 & q_0 \end{vmatrix}.
\end{aligned}$$ (B.13)

But, by virtue of Eqs. (3.26) and (3.43),

$$(1 - \rho_0 u)(\alpha B - A|\rho_0|M \operatorname{sinc} \theta)$$

$$= \frac{1}{n_0} [\alpha(\beta t + n_0 \cos \theta) - \beta(\alpha t - n_0 \cos(\psi_0 - \theta))]$$

$$= \alpha \cos \theta + \beta \cos(\psi_0 - \theta) = 1,$$ (B.14)

the last step being accomplished with the help of Eqs. (3.36). This verifies that

$$xq - yp = x_0 q_0 - y_0 p_0$$ (B.15)

holds at each point of a ray.

Appendix C
DERIVATION OF THIRD-ORDER TRACING FORMULAS FOR A RADIAL GRADIENT

When Eq. (6.27) is used on the right-hand side of Eq. (6.26) the first equation takes the form

$$\frac{d^2x}{d\bar{z}^2} + x = \hat{N}(x_0\bar{c} + y_0\bar{s})(\bar{c}^2 u + \bar{c}\bar{s}v + \bar{s}^2 w), \qquad (C.1)$$

where

$$\bar{c} = \cos\bar{z}, \qquad \bar{s} = \sin\bar{z}, \qquad \bar{z} = \frac{(z - z_0)b}{l_0}, \qquad (C.2)$$

and u, v, and w are defined by Eqs. (6.31). The solution of the corresponding homogeneous equation has the form

$$x_1 = A\bar{c} + B\bar{s}. \qquad (C.3)$$

Noting the trigonometric identities

$$\bar{c}^3 = \frac{c' + 3\bar{c}}{4},$$

$$\bar{s}^3 = \frac{3\bar{s} - s'}{4},$$

$$\bar{c}^2\bar{s} = \frac{\bar{s} + s'}{4}, \tag{C.4}$$

$$\bar{c}\bar{s}^2 = \frac{\bar{c} - c'}{4},$$

where

$$c' = \cos(3\bar{z}), \qquad s' = \sin(3\bar{z}), \tag{C.5}$$

we find, after a number of manipulations, that Eq. (C.1) can be written

$$\frac{d^2x}{d\bar{z}^2} + x = \frac{\hat{N}}{4} \{c'[x_0(u - w) - \bar{p}_0 v]$$
$$+ s'[x_0 v + \bar{p}_0(u - w)]$$
$$+ \bar{c}[x_0(3u + w) + \bar{p}_0 v]$$
$$+ \bar{s}[x_0 v + \bar{p}_0(u + 3w)]\}. \tag{C.6}$$

The next step is to look for a particular solution of the full equation in the form

$$X = \bar{a}_1 c' + \bar{a}_2 s' + \bar{z}(\bar{b}_1\bar{c} + \bar{b}_2\bar{s}), \tag{C.7}$$

where $\bar{a}_i$ and $\bar{b}_i$ are to be determined. This leads to

$$\frac{d^2X}{d\bar{z}^2} + X = \bar{a}_1 c' + \bar{a}_2 s' + \frac{1}{4}(\bar{b}_1\bar{s} - \bar{b}_2\bar{c}), \tag{C.8}$$

after which comparison with Eq. (C.6) gives

$$\overline{a}_1 = -\frac{\hat{N}}{32}\left[x_0(u - w) - \overline{p}_0 v\right],$$

$$\overline{a}_2 = -\frac{\hat{N}}{32}\left[x_0 v + \overline{p}_0(u - w)\right],$$

$$\overline{b}_1 = -\frac{\ddot{N}}{8}\left[x_0 v + \overline{p}_0(u + 3w)\right],$$

$$\overline{b}_2 = \frac{\hat{N}}{8}\left[x_0(3u + w) + \overline{p}_0 v\right]. \tag{C.9}$$

With X in Eq. (C.7) determined in this way, it is clear that the general solution of Eq. (C.1) has the form

$$x = A\overline{c} + B\overline{s} + \overline{a}_1 c' + \overline{a}_2 s' + \overline{z}(\overline{b}_1 \overline{c} + \overline{b}_2 \overline{s}), \tag{C.10}$$

where A and B are constants of integration. To find $\overline{p}$ we note that

$$\overline{p} = \frac{p}{b} = \frac{l\dot{x}}{b} = \frac{l_0 \dot{x}}{b} = \frac{dx}{d\overline{z}}. \tag{C.11}$$

Therefore we find, from Eq. (C.10),

$$\overline{p} = -A\overline{s} + B\overline{c} - 3\overline{a}_1 s' + 3\overline{a}_2 c' + \overline{b}_1 \overline{c} + \overline{b}_2 \overline{s} + \overline{z}(-\overline{b}_1 \overline{s} + \overline{b}_2 \overline{c}). \tag{C.12}$$

The constants A and B can now be determined from the initial conditions, i.e.,

$$x = x_0, \quad \overline{p} = \overline{p}_0 \quad \text{when} \quad \overline{z} = 0, \quad s = 0, \quad c = 1. \tag{C.13}$$

This leads to

$$A = x_0 - \overline{a}_1, \qquad B = \overline{p}_0 - 3\overline{a}_2 - \overline{b}_1. \tag{C.14}$$

Inserting these values and again making use of the identities in Eq. (C.4) leads to

$$x = \overline{c}x_0 + \overline{s}\overline{p}_0 - 4\overline{a}_1 \overline{c}s^2 - 4\overline{a}_2 \overline{s}^3 + \overline{b}_1(-\overline{s} + \overline{c}\overline{z}) + \overline{b}_2 \overline{s}\overline{z} \tag{C.15}$$

and

$$\overline{p} = \overline{c}p_0 - \overline{s}x_0 + 4\overline{a}_1\overline{s}(1 - 3\overline{c}^2) - 12\overline{a}_2\overline{c}s^2 - \overline{b}_1\overline{s}z + \overline{b}_2(\overline{s} + \overline{c}z). \quad \text{(C.16)}$$

It is convenient to remove a scale factor from the constants $\overline{a}_i$ and $\overline{b}_i$, so that in place of Eqs. (C.9) we have

$$\begin{aligned}
a_1 &= x_0(u - w) - \overline{p}_0 v, \\
a_2 &= x_0 v + \overline{p}_0(u - w), \\
b_1 &= x_0 v + \overline{p}_0(u + 3w), \\
b_2 &= x_0(3u + w) + \overline{p}_0 v).
\end{aligned} \quad \text{(C.17)}$$

Then we have

$$x = \overline{c}x_0 + \overline{s}p_0 + \frac{\hat{N}}{8}\left[a_1\overline{c}s^2 + a_2\overline{s}^3 + b_1(\overline{s} - \overline{c}z) + b_2\overline{s}z\right] \quad \text{(C.18)}$$

and

$$\overline{p} = \overline{c}p_0 - \overline{s}x_0 + \frac{\hat{N}}{8}\left[a_1(3\overline{c}^2 - 1) + 3a_2\overline{c}s^2 + b_1\overline{s}z + b_2(\overline{s} + \overline{c}z)\right]. \quad \text{(C.19)}$$

With a view to using these formulas for evaluating third-order aberrations, it is important to separate the paraxial and higher-order terms in the variables x_0, y_0, $\overline{p}_0$, and $\overline{q}_0$. For this purpose we must recall Eq. (C.2) and note, with the help of Eq. (6.20), that

$$\begin{aligned}
l_0 &= (n_0^2 - p_0^2 - q_0^2)^{1/2} \\
&= \left[N_0^2(1 + 2\overline{N}_1 r_0^2) - p_0^2 - q_0^2\right]^{1/2} \\
&= N_0\left[1 + \overline{N}_1 r_0^2 - \frac{1}{2}\frac{(\overline{p}_0^2 + \overline{q}_0^2)b^2}{N_0^2}\right].
\end{aligned} \quad \text{(C.20)}$$

But, with $N_1 < 0$, Eq. (6.23) gives

$$\frac{\frac{1}{2}b^2}{N_0^2} = -\overline{N}_1, \quad \text{(C.21)}$$

so that

$$l_0 = N_0(1 + W), \quad \text{(C.22)}$$

where Eqs. (6.31) and (6.33) were used. Evidently W is of second order in the ray variables.

Using this result we have, if $z_0 = 0$,

$$\bar{z} = \frac{b}{N_0} z(1 - W) = kz(1 - W). \tag{C.23}$$

and, with W small,

$$\bar{c} = \cos\bar{z} = \cos(kz) + kzW\sin(kz),$$
$$\bar{s} = \sin\bar{z} = \sin(kz) - kzW\cos(kz). \tag{C.24}$$

We now substitute these results into Eqs. (C.18) and (C.19), where the terms involving W can be dropped in the last four terms of each formula since the a_i and b_i are already of third order. Writing now

$$c = \cos(kz), \qquad s = \sin(kz), \tag{C.25}$$

we have then

$$x = cx_0 + s\bar{p}_0 + kzW(sx_0 - c\bar{p}_0)$$
$$+ \frac{\hat{N}}{8}\left[a_1 cs^2 + a_2 s^3 + b_1(s - ckz) + b_2 skz\right] \tag{C.26}$$

and

$$\bar{p} = c\bar{p}_0 - sx_0 + kzW(s\bar{p}_0 + cx_0)$$
$$+ \frac{\hat{N}}{8}\left[a_1 s(3c^2 - 1) + 3a_2 cs^2 + b_1 skz + b_2(s + ckz)\right]. \tag{C.27}$$

Thus, Eqs. (6.32), (6.34), and (6.35) have been established. When $z_0 \neq 0$, z must be replaced by $z - z_0$ in the above equations.

Appendix D

DERIVATION OF EQUATIONS (10.10) TO (10.16)

To derive the third-order aberration formulas for the special axial-gradient lens we first obtain the third-order tracing formulas for this lens. The lens is pictured in Fig. 10.1, where the meaning of the various symbols is displayed. The entire strategy for deriving the third-order aberration formulas for this lens is very similar to that used in Chapter 8 for the Wood lens.

At the terminal point of a ray, we have, from Eq. (10.1), in the third-order approximation,

$$n_1 = N_0(1 + \alpha z_1)^{1/2} = N_0[1 + \alpha(d + \bar{z}_1)]^{1/2} = N_1(1 + \hat{\alpha}\bar{z}_1) \quad \text{(D.1)}$$

with

$$a = (1 + \alpha d)^{1/2}, \qquad N_1 = N_0 a, \qquad \hat{\alpha} = \frac{1}{2}\alpha/a^2. \quad \text{(D.2)}$$

Also, from the equation of the sphere,

$$\bar{z}_1 = \frac{cy_1^2}{[1 + (1 - c^2 y_1^2)^{1/2}]} \cong \frac{cy_1^2}{2}, \quad \text{(D.3)}$$

where c is the curvature of the sphere.

With the help of Eq. (10.3) we find, for a ray in the yz-plane,

$$l_0 = N_0\left(1 - \frac{\bar{q}_0^2}{2}\right),$$

$$l_1 = N_1\left[1 + \hat{\alpha}\bar{z}_1 - \frac{\bar{\bar{q}}_0^2}{2}\right],$$

(D.4)

where

$$\bar{q}_0 = \frac{q_0}{N_0}, \qquad \bar{\bar{q}}_0 = \frac{\bar{q}_0}{a}.$$

(D.5)

Using these results we find the height and direction of the ray as it strikes the second surface, i.e.,

$$q_1 = q_0,$$
$$y_1 = \bar{y} + \bar{\bar{q}}_0(\bar{z}_1 + \bar{q}_0^2\bar{d}/2),$$

(D.6)

where

$$\bar{y} = y_0 + \bar{d}q_0, \qquad \bar{d} = \frac{2d}{a+1}.$$

(D.7)

Refraction at the curved surface is conveniently expressed vectorially by

$$\mathbf{s}' = \mathbf{s}_1 + \Gamma\mathbf{N},$$

(D.8)

where $\mathbf{s}_1$ and $\mathbf{s}'$ are the optical direction vectors of lengths n_1 and unity respectively, and $\mathbf{N}$ is the unit normal vector to the sphere. By differentiation of the equation of the sphere we find

$$\mathbf{N} = (-cx_1, -cy_1, 1 - c\bar{z}_1).$$

(D.9)

It follows that

$$\mathbf{N} \cdot \mathbf{s}_1 = -cy_1q_0 + (1 - c\bar{z}_1)l_1$$

$$= N_1\left[1 + \hat{\alpha}\bar{z}_1 - \frac{1}{2}(c\bar{y} + \bar{q}_0)^2\right],$$

(D.10)

which is needed to determine Γ.

From Eq. (D.8) we find

$$\mathbf{s}' \cdot \mathbf{s}_1 = n_1^2 + \Gamma(\mathbf{N} \cdot \mathbf{s}_1),$$
$$\Gamma^2 = (\mathbf{s}' - \mathbf{s}_1)^2 = 1 - 2\mathbf{s}' \cdot \mathbf{s}_1 + n_1^2. \tag{D.11}$$

Elimination of $\mathbf{s}' \cdot \mathbf{s}_1$ then leads to

$$\Gamma = [1 + (\mathbf{N} \cdot \mathbf{s}_1)^2 - n_1^2]^{1/2} - \mathbf{N} \cdot \mathbf{s}_1, \tag{D.12}$$

the positive root being chosen so that when $n_1 \to 1$, $\Gamma \to 0$ under the assumption that $\mathbf{N} \cdot \mathbf{s}_1 > 0$.

Before completing the third-order refraction analysis we consider the paraxial approximation and thereby determine the focal length and back-focus. To define the focal length of the lens we investigate a paraxial ray in the plane $x = 0$ and note, from Eqs. (D.1), (D.7), and (D.10), that in this approximation

$$q_1 = q_0, \quad y = \bar{y}, \quad y_0 = \overline{d q_0}, \quad n_1 = N_1 = N_0 a, \quad \mathbf{N} \cdot \mathbf{s}_1 = N_1. \tag{D.13}$$

Therefore, from Eq. (D.12),

$$\Gamma = 1 - N_1 \tag{D.14}$$

and the second component of Eq. (D.8) gives

$$q' = q_0 - c\bar{y}\Gamma = q_0 - \frac{c\bar{y}}{1 - N_1} \tag{D.15}$$

in this approximation.

Now, specializing to a ray with $q_0 = 0$ and $y_0 = h$, we define the focal length f by

$$\frac{1}{f} = \lim_{h \to 0} \frac{\tan \sigma'}{h} = -\lim_{h \to 0} \frac{q'}{h}, \tag{D.16}$$

for this ray. Hence

$$\frac{1}{f} = c1' = c(1 - N_1). \tag{D.17}$$

The back-focus f' is defined by

$$\frac{1}{f'} = -\lim_{h \to 0} \frac{q'}{y_1} \tag{D.18}$$

for the same ray as above.

It follows that $f' = f$ in this example. Thus Eqs. (10.15) are verified.

The third-order refraction is accomplished by substituting Eqs. (D.1) and (D.10) into Eq. (D.12). By noting Eq. (10.15), the result can be written

$$\Gamma = \frac{1}{cf}\left\{ 1 + N_1\left[\frac{1}{2}(\bar{\bar{q}}_0 + c\bar{y})^2 - cf\alpha\bar{z}_1 \right] \right\}. \tag{D.19}$$

The second component of Eq. (D.8) gives

$$q' = q_0 - cy_1\Gamma, \tag{D.20}$$

after which it follows from Eqs. (D.6) and (D.19) that

$$q' = q_0 - \frac{y}{f} - \frac{\bar{\bar{q}}_0}{f}\left(\bar{z}_1 + \frac{\bar{q}_0^2\bar{d}}{2} \right)$$
$$+ \frac{N_1\bar{y}}{f}\left[cf\hat{\alpha}\bar{z}_1 - \frac{1}{2}(\bar{\bar{q}}_0 + c\bar{y})^2 \right]. \tag{D.21}$$

Thus the direction of the ray in the image region is determined.

The ray height in the focal plane (see Fig. 10.1) is determined by the relation

$$y' - y_1 = (f' - \bar{z}_1)\tan(-\sigma'), \tag{D.22}$$

or

$$y' = y_1 + f'\left(q' + \frac{1}{2}q'^3 \right) - q'\bar{z}_1 \tag{D.23}$$

to third order. Inserting the third-order formulas for y_1 and q' given above leads to

$$y' = q_0 f + \frac{1}{2}f\left(q_0 - \frac{\bar{y}}{f} \right)^3 - \bar{z}_1\left(q_0 - \frac{\bar{y}}{f} \right)$$
$$+ N_1\bar{y}\left[cf\hat{\alpha}\bar{z}_1 - \frac{1}{2}(\bar{\bar{q}}_0 + c\bar{y})^2 \right]. \tag{D.24}$$

The first term here shows that the paraxial ray height is given by

$$y_0' = q_0 f, \tag{D.25}$$

which verifies Eq. (10.16) and shows that all paraxial rays in the plane $x = 0$ and starting at a given object point at infinity converge to the same point in the focal plane. Actually it can be shown that *all* paraxial rays from a given object point converge to a single image point.

The lateral spherical aberration $\overline{S}$ of the lens is found by considering Eq. (D.24) applied to a ray with $q_0 = 0$, $y_0 = h$, $\overline{y} = h$, and $\overline{z}_1 = ch^2/2$, in which case y' gives the value of $\overline{S}$. The result is

$$\overline{S} = \frac{1}{2} N_1 c^2 h^3 (\hat{a}f - N_1). \tag{D.26}$$

The longitudinal spherical aberration is then easily deduced from this by noting that

$$\frac{\overline{S}}{S} = \frac{h}{f} \tag{D.27}$$

for the same ray considered above. This verifies Eq. (10.10).

To derive a formula for coma we recall the definition given by Eq. (7.1) and refer to Fig. 7.1, where now we assume the Wood lens pictured there replaced by the axial-gradient lens presently being considered. Applying Eq. (D. 24) to each of the three rays and inserting the results in Eq. (7.1) leads to

$$C = \frac{-3}{2} N_1 c \rho^2 \left[q_0 - \frac{\overline{y}}{f} + c\overline{y}(1 - f\hat{a}) \right]. \tag{D.28}$$

In the paraxial approximation, which can be applied to factors here, we have, from Fig. 10.1,

$$y_0 = -z_e \sin(-\sigma) = -z_e q_0. \tag{D.29}$$

Hence, from Eq. (D.7),

$$\overline{y} = -z_e q_0 + \frac{\overline{d}q_0}{N_0} = -q_0 \overline{z}_e, \tag{D.30}$$

where

$$\overline{z}_e = z_e - \frac{\overline{d}}{N_0}. \tag{D.31}$$

Substitution for $\bar{y}$ in Eq. (D.28) then leads to

$$C = \frac{-3}{2} N_1 c q_0 \rho^2 \left[1 + \frac{\bar{z}_e}{f} - c\bar{z}_e(1 - f\hat{\alpha}) \right], \qquad (D.32)$$

which verifies Eq. (10.11).

The distortion formula is derived by applying Eq. (D.24) to the definition given by Eq. (7.2). Thus we have

$$D = \frac{50}{q_0 f} \left\{ f\left(q_0 - \frac{\bar{y}}{f} \right)^3 - 2\bar{z}_1\left(q_0 - \frac{\bar{y}}{f} \right) \right.$$

$$\left. + N_1 \bar{y}[cf\hat{\alpha}\bar{z}_1 - (\bar{\bar{q}}_0 + c\bar{y})^2] \right\}. \qquad (D.33)$$

Noting that, here again, we can apply the paraxial formula, Eq. (D.30), and the relation

$$\bar{z}_1 = \frac{1}{2} c\bar{y}^2 = \frac{1}{2} cq_0^2\bar{z}_e^2, \qquad (D.34)$$

we are able to verify Eq. (10.14).

The derivations of the formulas for the astigmatism and Petzval curvature depend on first obtaining those for the meridional and sagittal curvatures Z_m and Z_s respectively.

For Z_m the definition given by Eq. (7.3) is applied, taking the form

$$Z_m = -f \lim_{h \to 0} \frac{y' - y'_p}{h} \qquad (D.35)$$

for the two rays pictured in Fig. 7.2. Using Eq. (D.24) for each ray and again taking account of Eqs. (D.30) and (D.31), we are led to the result

$$Z_m = \frac{3}{2} f q_0^2 \left[\left(1 + \frac{\bar{z}_e}{f} \right)^2 - c\bar{z}_e\left(2 + \frac{\bar{z}_e}{f} \right) \right.$$

$$\left. + \frac{1}{3N_1} + N_1 c^2\bar{z}^2(1 - f\hat{\alpha}) \right]. \qquad (D.36)$$

The sagittal analog of Eq. (D.35) is found to be

$$Z_s = \lim_{h' \to 0} \left[f' - \bar{z}_1 + \frac{x_1 l'}{p'} \right], \qquad (D.37)$$

applied to the skew ray pictured in Fig. 7.3. In this case, from Eqs. (D.8) and (D.9),

$$x_1 = h', \qquad p' = -ch'\Gamma, \qquad l' = l_1 + \Gamma(1 - c\bar{z}_1), \qquad \text{(D.38)}$$

with the result that

$$Z_{\mathrm{s}} = f - \frac{1 + l_1/\Gamma}{c}. \qquad \text{(D.39)}$$

Here the limiting values of l_1 and Γ are given by Eqs. (D.4) and (D.19) respectively, leading to

$$Z_{\mathrm{s}} = N_1 f\left[\frac{1}{2} N_1(c\bar{y} + \bar{\bar{q}}_0)^2 + \frac{1}{2}\bar{\bar{q}}_0^2 - cf\hat{\alpha}\bar{z}_1\right]. \qquad \text{(D.40)}$$

Taking account of Eq. (D.30) as before, we find

$$Z_{\mathrm{s}} = \frac{1}{2} fq_0^2\left[\frac{1}{N_1} + \left(1 + \frac{\bar{z}_{\mathrm{e}}}{f}\right)^2 - c\bar{z}_{\mathrm{e}}\left(2 + \frac{\bar{z}_{\mathrm{e}}}{f}\right)\right.$$
$$\left. + N_1 c^2\bar{z}_{\mathrm{e}}^2(1 - \hat{\alpha}f)\right]. \qquad \text{(D.41)}$$

It is now a simple matter to obtain the astigmatism and Petzval curvature formulas by noting the definitions given by Eqs. (7.5) and (7.6), i.e.,

$$A = \frac{Z_{\mathrm{m}} - Z_{\mathrm{s}}}{2}, \qquad P = Z_{\mathrm{m}} - 3Z_{\mathrm{s}}. \qquad \text{(D.42)}$$

Thus, Eqs. (10.13) and (10.14) are verified.

REFERENCES

Abelès, F. (1950). *Ann. Phys. (Paris)* **5**, 596.

Berreman, D. W. (1956). *Bell Syst. Tech. J.* **44**, 2117.

Born, M., and E. Wolf (1970). "Principles of Optics," 4th ed. Pergamon, Oxford.

Brinkmeyer, E. (1977). *Appl. Opt.* **16**, 2802.

Brushon, J. (1975). *Opt. Acta* **22**, 211.

Buchdahl, H. (1969). "Optical Aberration Coefficients." Dover, New York.

Buchdahl, H. (1970). "An Introduction to Hamiltonian Optics." Cambridge Univ. Press, London and New York.

Cheng, Y. C., W. D. Westwood, E. A. Sullivan, S. J. Ingrey, and J. Fox (1975). *Proc. Symp. Opt. and Acoustical Microelectronics, Brooklyn, 1975*, p. 271. Polytech. Press, Brooklyn, New York.

Cohen, L. G. (1976). *Appl. Opt.* **15**, 1808.

Eaton, J. E. (1953). U.S. Naval Research Lab Report No. 4110.

Fletcher, A., T. Murphy, and A. Young (1954). *Proc. Roy. Soc. (London)* **223**, 216.

Foelster, K., and R. Herrmann (1974). *Opt. Acta* **21**, 25.

French, W. G., G. W. Tasker, and J. R. Simpson (1976). *Appl. Opt.* **15**, 1803.

Gloge, D., and E. A. J. Marcatili (1973). *Bell Syst. Tech. J.* **52**, 1563.

Gupta, A., K. Thyagarajan, I. C. Goyal, and A. K. Ghatak (1976). *J. Opt. Soc. Amer.* **66**, 1320.

Hamblen, D. P. (1969). U.S. Patent No. 3,486,808.

Hamblen, D. P. (1977). U.S. Patent No. 4,022,855.

Herzberger, M. (1958). "Modern Geometrical Optics," p. 426. Wiley (Interscience), New York.

Houghton, A. J., and P. D. Townsend (1976). *Appl. Phys. Lett.* **29**, 565.

Huynen, J. R. (1958). *IRE WESCON Conv. Rec., Part 1*, 219.

Iga, K., and N. Yamamoto (1977). *Appl. Opt.* **16**, 1305.

Iga, K., K. Yokomori, and T. Sakayori (1975). *Appl. Phys. Lett.* **26**, 578.

Ikeda, M., M. Tateda, and H. Yoshikiyo (1975). *Appl. Opt.* **14**, 814.

Jacobsson, R. (1963). *In* "Progress in Optics" (E. Wolf, ed.), Vol. V. North–Holland Publ., Amsterdam, p. 249.

Jacobsson, R. (1968). *In* "Optical Properties of Dielectric Films" (N. N. Axelrod, ed.). Electrochemical Society, New York.

Kapron, F. P. (1970). *J. Opt. Soc. Amer.* **60**, 1433.

Kawakami, S., and J. Nishizawa (1968). *IEEE Trans. Microwave Theory Tech.* **16**, 814.

Kita, H., and T. Uchida (1970). *Opt. Spectra* **4**, 80.

Kitano, I., K. Koizumi, and Y. Ikeda (1972). U. S. Patent No. 3,830,640.

Kornhauser, E. T., and A. D. Yaghjian (1967). *Radio Sci.* **2**, 299.

Luneburg, R. K. (1964). "Mathematical Theory of Optics." Univ. of California Press, Berkeley. Previously publ. Brown Univ., 1944.

Maeda, K., and J. Hamasaki (1977). *J. Opt. Soc. Amer.* **67**, 1672.

Mahan, A. I. (1962). *Appl. Opt.* **1**, 497.

Marcatili, E. A. J. (1967). *Bell Syst. Tech. J.* **46**, 149.

Marchand, E. W. (1970). *J. Opt. Soc. Amer.* **60**, 1.

Marchand, E. W. (1972). *Appl. Opt.* **11**, 1104.

Marchand, E. W. (1973). *In* "Progress in Optics" (E. Wolf, ed.), Vol. XI, p. 305. North-Holland Publ., Amsterdam.

Marchand, E. W. (1976). *J. Opt. Soc. Amer.* **66**, 1326.

Marchand, E. W., and D. J. Janeczko (1974). *J. Opt. Soc. Amer.* **64**, 846.

Marcuse, D. (1965a). *IEEE Trans. Microwave Theory Tech.* **13**, 11.

Marcuse, D. (1965b). *Bell Syst. Tech. J.* **44**, 2065.

Marcuse, D., and S. E. Miller (1964). *Bell Syst. Tech. J.* **43**, 1759.

Marcuse, D., and H. M. Presby (1975). *J. Opt. Soc. Amer.* **65**, 367.

Marhic, M. E., P. S. Ho, and M. Epstein (1975). *Appl. Phys. Lett.* **26**, 574.

Martin, W. E. (1974). *Appl. Opt.* **13**, 2112.

Martin, W. E. (1975). *Appl. Opt.* **14**, 2427.

Matsumura, H., S. Yoshikawa, and R. Tatsumi (1972). U. S. Patent No. 3,637,295.

Matsushita, K., K. Nishizawa, and M. Toyama (1972). U. S. Patent No. 3,827,785.

Maxwell, J. C. (1854). *Cambridge and Dublin Math. J.* **8**, 188. [Also *Scientific Papers*, I, p. 76, Cambridge Univ. Press, London and New York.]

Miller, S. E. (1965). *Bell Syst. Tech. J.* **44**, 2013.

Montagnino, L. (1968). *J. Opt. Soc. Amer.* **58**, 1667.

Moore, D. T. (1971). *J. Opt. Soc. Amer.* **61**, 886.

Moore, D. T. (1975), *J. Opt. Soc. Amer.* **65**, 451.

Moore, D. T. (1977a). *J. Opt. Soc. Amer.* **67**, 1137.

Moore, D. T. (1977b). *J. Opt. Soc. Amer.* **67**, 1143.

Moore, D. T., and P. J. Sands (1971). *J. Opt. Soc. Amer.* **61**, 1195.

Moore, D. T., and P. J. Sands (1973). U. S. Patent No. 3,729,253.

Moore, R. (1971). U.S. Patent No. 3,816,160.

Moore, R. (1973). U.S. Patent No. 3,718,383.

Morgan, S. P. (1958), *J. Appl. Phys.* **29**, 1358.

Nishijima, Y., and G. Oster (1964). *J. Opt. Soc. Amer.* **54**, 1.

O'Connor, P. B., J. B. Macchesney, H. M. Presby, and L. G. Cohen (1976). *Amer. Ceram. Soc. Bull.* **55**, 513–16, 519.

Ohtsuka, Y. (1973). *Appl. Phys. Lett.* **23**, 247.

Ohtsuka, Y., and Y. Shimizu (1977). *Appl. Opt.* **16**, 1051.

Okashi, T., and K. Hotate (1976). *Appl. Opt.* **15**, 2756.

Paxton, K. B., and W. Streifer (1971a). *Appl. Opt.* **10**, 769. [See also p. 1164.]

Paxton, K. B., and W. Streifer (1971b). *Appl. Opt.* **10**, 2090.

Pearson, A. D., W. G. French, and E. G. Rawson (1969). *Appl. Phys. Lett.* **15**, 76.

Presby, H. M., (1974). *J. Opt. Soc. Amer.* **64**, 280

Rawson, E. G., D. R. Herriott, and J. McKenna (1970). *Appl. Opt.* **9**, 753.

Ritchie, I. T., and B. Window (1977). *Appl. Opt.* **16**, 1438.

Sands, P. J. (1970). *J. Opt. Soc. Amer.* **60**, 1436.

Sands, P. J. (1971a). *J. Opt. Soc. Amer.* **61**, 777.

Sands, P. J. (1971b). *J. Opt. Soc. Amer.* **61**, 879.

Sands, P. J. (1971c). *J. Opt. Soc. Amer.* **61**, 1086.

Sands, P. J. (1971d). *J. Opt. Soc. Amer.* **61**, 1495.

Sands, P. J. (1973). *J. Opt. Soc. Amer.* **63**, 1210.

Saunders, M. J., and W. B. Gardner (1977). *Appl. Opt.* **16**, 2369.

Schulze, A. G. (1913). *Ann. Phys.* **40**, 335.

Schulze, A. G. (1915). *Ann. Physik. Chem.* **89**, 168.

Schulze, A. G. (1919). *Z. Elektrochem.* **19**, 122.

Sinai, P. (1971). *Appl. Opt.* **10**, 99.

Sommer, R. G., R. D. Deluca, and G. E. Burke (1976). *Electron. Lett.* **12**, 408.

Southwell, W. H. (1977a). *J. Opt. Soc. Amer.* **67**, 1004.

Southwell, W. H. (1977b). *J. Opt. Soc. Amer.* **67**, 1010.

Stettler, R. (1955). *Optik* **12**, 529.

Stone, J., and C. A. Burrus (1975). *Appl. Opt.* **14**, 151.

Stone, F. T. (1977). *Appl. Opt.* **16**, 2738.

Streifer, W., and C. N. Kurtz (1967). *J. Opt. Soc. Amer.* **57**, 779.

Toraldo di Francia, G. (1957). *Ann. Mat. Pura* **44**, 35.

Uslenghi, P. L. E. (1964). *Can. J. Phys.* **42**, 2359.

Valette, S., G. Labrunie, J. Deutsch, and J. Lizet (1977). *Appl. Opt.* **16**, 1289.

Van Ass, H. M. J. M., R. G. Gossink, and P.S.W. Severin (1976). *Electron. Lett.* **15**, 369.

Watkins, L. S. (1974). *J. Opt. Soc. Amer.* **64**, 767.

White, R. (1975). *J. Opt. Soc. Amer.* **65**, 676.

Wonsiewicz, B. C., W. G. French, P. D. Lazay, and J. R. Simpson (1976). *Appl. Opt.* **15**, 1048.

Wood, R. W. (1905). "Physical Optics," p. 71. Macmillan, New York.

Zachariasen, W. H. (1932a). *J. Am. Chem. Soc.* **54**, 3841.

Zachariasen, W. H. (1932b), *Phys. Rev.* **39**, 185.

INDEX

A
B
C 8
D 9
E 0
F 1
G 2
H 3
I 4
J 5